ALEXCHOI design & Partners

collections
蔡明治设计精选作品集

Volume One
Adventure
第一册 破旧立新

深圳市艺力文化发展有限公司 编

华南理工大学出版社
SOUTH CHINA UNIVERSITY OF TECHNOLOGY PRESS
·广州·

Everything has exception, all you need to do is step forward.... We shared common beliefs of creativity, simplicity & details. And the synergy of beliefs has opened up a brighter future for other local designers in all kind of industry.

SINCE 1997

ADVENTURE

———————————————————————————————— SINCE 1997

图书在版编目（CIP）数据

蔡明治设计精选作品集 = ALEXCHOI design & partners collections.1，破旧立新 / 深圳市艺力文化发展有限公司编．— 广州：华南理工大学出版社，2014.4
 ISBN 978-7-5623-4118-5

Ⅰ．①蔡… Ⅱ．①深… Ⅲ．①室内装饰设计－作品集－中国－现代 Ⅳ．① TU238

中国版本图书馆 CIP 数据核字（2014）第 059403 号

蔡明治设计精选作品集 ALEXCHOI design & Partners collections
第一册 破旧立新 Volume One Adventure
深圳市艺力文化发展有限公司 编

出 版 人：**韩中伟**
出版发行：华南理工大学出版社
 （广州五山华南理工大学 17 号楼，邮编 510640）
 http://www.scutpress.com.cn E-mail: scutc13@scut.edu.cn
 营销部电话：020-87113487 87111048（传真）
策划编辑：赖淑华
责任编辑：陈 昊 赖淑华
印 刷 者：深圳市汇亿丰印刷包装有限公司
开 本：787mm×1092mm 1/16 印张：22.75
成品尺寸：185mm × 260mm
版 次：2014 年 4 月第 1 版 2014 年 4 月第 1 次印刷
定 价：398.00 元（共 3 册）

版权所有 盗版必究 印装差错 负责调换

CONTENTS

RETAIL
零售

Azona a02 flagship store Azona a02 旗舰店	2
F.C.K flagship store F.C.K. 服饰旗舰店	8
BMW showroom 宝马汽车陈列中心	10
Semir chain store 森马服饰连锁店	16
K-Boxing men's fashion concept store 劲霸男士服饰概念店	26
Xplus shopping hub Xplus 购物中心	32

OFFICE
办公室

KML Engineering Ltd. office 高明科技工程有限公司办公室	40
Global Star Entertainment & Technology office Global Star Entertainment & Technology 办公室	46
Studios of Centro Digital Pictures 先涛数码影画制作有限公司工作室	52
C.F.L. Enterprise Limited office 赛辉洋行有限公司办公室	58
Artapower International Group Ltd. office 艺达堡集团有限公司办公室	66
Pearltower Garments & Toys Co. Ltd. office 宝台制衣玩具有限公司办公室	70
Artapower International Group Ltd. headquarter 艺达堡集团有限公司总部办公室	76
Jeanswest International (H.K.) Ltd. - China head office 真维斯国际(香港)有限公司——中国总部办公室	80

EXHIBITION & SHOWFLAT
展览厅

Young Achievers' Gallery 香港教育局荟萃馆	86
CLP Energy Efficiency Exhibition Centre 中华电力能源效益展览中心	92
CLP Smart Grid Experience Centre 中华电力智能电网体验馆	96

RESIDENTIAL
住宅

A Private Villa in Santa Monica, L.A. 洛杉矶圣塔莫尼卡私人别墅	102
Bachelor's flat in Kowloon Tong 九龙塘住宅项目	112

RETAIL
零售

Azona a02 flagship store
Azona a02 旗舰店

F.C.K Flagship Store
F.C.K. 服饰旗舰店

BMW showroom
宝马汽车陈列中心

Semir chain store
森马服饰连锁店

K-Boxing men's fashion concept store
劲霸男士服饰概念店

Xplus shopping hub
Xplus 购物中心

Azona a02 flagship store

Azona a02 旗舰店

Hong Kong, China
中国香港

Area : 278.71m²
Completed Year : 2001
Design : Alex Choi
Photographer : ALEXCHOI design & Partners

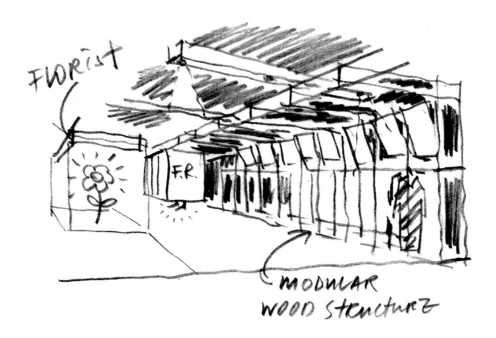

Azona a02 is a quirky, playful "lifestyle" shop that carries a rich mix of clothes, accessories, music CDs and magazines, with the balance and range of items on display varying from shop to shop depending on location and demand. Asked to come up with a new "look" for the company's existing and future stores, the designers therefore needed to devise a flexible system that would allow the creation of numerous outlets, each with its own rich layering of spaces but all finished to the same high quality. The designers chose to assign each product type a designated "zone", creating a shop-in-shop configuration, while maintaining a cohesive look for the space as a whole through the consistent use of simple wooden oak frames. In this way, the shop maintains visual transparency, where customers are afforded views across different zones, allowing the space to overlap and disseminate. The wooden frames take on a chameleon-like quality, appearing as wall linings, ceilings and even freestanding or hanging racks, depending on different site situations. In the 557.41 m² Causeway Bay shop, for example, a wooden frame acts as a ceiling, delineating the glass enclosure of a small flower stall. Elsewhere, they act as display systems to showcase the clothes and shoes. Steps in floor height and different floor finishes are combined with the careful positioning of island display units to help guide the customers through the different areas of the shop, with many of the products on open display, encouraging customers to browse, touch and play. An extensive use of wooden flooring adds domestic warmth and friendliness to the overall ambience.

Azona a02 是一间别具个性的生活概念连锁店,集服务、配饰、音乐和杂志于一体,每间分店由于地点及顾客要求不同,陈列的产品种类亦大相径庭。客户要求设计师现有及即将开设的分店构思一个全新形象,设计必须极具灵活性,以配合不同的店铺环境,同时保留各自的空间层次感,确保设计质素保持一致。设计师遂为各款产品赋予独立的展示区,营造出店中有店的布局,并将橡木框架重复运用到各分店中,令品牌的形象更为统一。如此一来,店内虽划分成不同的产品区域,却保持了视线的连系性,令空间可以相互重叠及渗透。

橡木框架恰如变色龙般,以不同形式出现在空间之中,如墙饰、天花,甚至是独立的挂衣货架,视店铺环境而定。以面积达 557.4㎡ 的铜锣湾店为例,橡木框架天花围合出一个小小的玻璃花店,至于其他地方则用作陈列服装及鞋履的货架。设计师利用梯级地台和不同的地板物料,勾勒出一条清晰的购物路线,加上货岛的巧妙配合,引领顾客穿梭于店内的不同区域。大部分产品均采用互动开放的陈列方式,鼓励客户拿在手中慢慢欣赏,配合柔和的木色地板,令客人倍感亲切自在。

2001 | Azona a02 flagship store

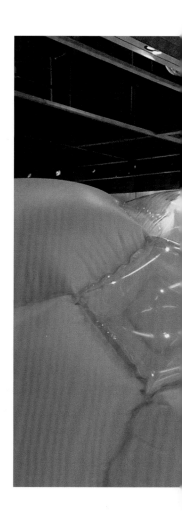

Azona a02 flagship store

F.C.K flagship store
F.C.K. 服饰旗舰店

Hong Kong, China
中国香港

Completion Year : 1997
Design : ALex Choi
Photographer : ALEXCHOI design & Partners

1997 | F.C.K flagship store

Back to 90's, Hong Kong fashion industry was stronger then ever, many up and coming and famous fashion labels from worldwide had chosen Hong Kong as a startup location for the Asia market. At that time, there were not many local fashion brands in Hong Kong and not mentioning to join this worldwide fashion battle. But anything has its exception, all you need to do is step forward.

F.C.K is one of the earliest local young fashion brands that flaunts the banner of "Made in Hong Kong" fashion. To stand out from a bunch of renowned International brands, the design went for the minimalism approach instead of embellishment. The design exaggerated the simplicity and brightness of the area so as to make it more spacious than its size. The two local designers who came from different background: Fashion and interior shared common beliefs of creativity, simplicity and detail, and the synergy of the beliefs has opened up a brighter future for Hong Kong designers in all kinds of industry.

追溯到 20 世纪 90 年代，香港时装行业是比以前任何时候都强，许多来自世界各地的未来时尚品牌和当时已经著名的时尚品牌都选择香港作为亚洲市场的启动位置。当时，香港没有许多本地的时尚品牌，也没有涉及加入全球时尚品牌的竞争中。但任何事情都有例外，你所需要做的是向前迈进一步。

F.C.K 是当地最早的年轻时尚品牌之一，招展着"香港制造"时装的旗帜。为了从一群知名国际品牌中脱颖而出，设计运用了简约主义的方法而不是大量的装饰。设计放大区域的简约性和明亮感，以使它看起来比其实际大小更宽敞。两个来自不同背景的本地设计师：时装设计师和室内设计师共同信仰的创造力、简约、细节以及协作，为各行各业的香港设计师开辟了一个光明的未来。

BMW showroom
宝马汽车陈列中心

Xiamen, China
中国厦门

Area : 1,393.55m²
Completion Year : 2003
Design : Alex Choi
Photographer : AlEXCHOI design & Partners

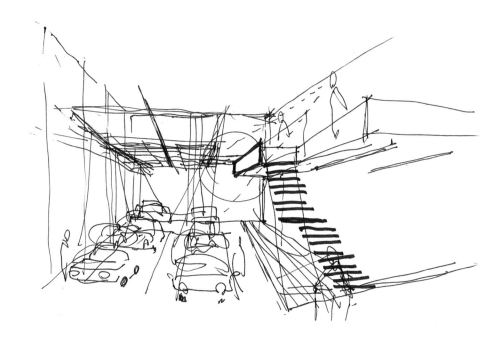

The significance of this project lies in the fact that this was not only the first flagship BMW showroom in Xiamen, but also the first of its kind to break away from the global design guidelines laid down by the headquarters in Germany. Rather than the standard, sparklingly clean white box in which the cars are displayed almost as works of art, the designers designed the showroom with a new look fashioned primarily in glass and metal, turning the whole space into a display of the three-dimensional design excellence for which BMW is renowned. The scale of the space was impressive, with a generous ceiling height that allowed designers to insert a mezzanine floor to house the sales offices. A linear staircase leads up to a broad balcony, from where customers can appreciate a bird's eye view of the car they might be about to order. Light in weight and slender of proportion, the staircase was conceived as the showroom's focal point, commanding the immediate attention of all who enter. The existing steel soffit was left deliberately exposed, but partially concealed by localised ceiling panels and suspended lighting frames that follow the linear geometry of the architecture. This was also reflected in the way the vehicles are laid out, creating a streamlined look consistent with the rational precision central to German aesthetics.

本个案不仅是宝马位于厦门的首间旗舰店，亦是全球首个率先打破德国总部既定设计规范的陈列室，故此别具意义。有别于一贯标准的白色方块空间，以及纯粹视轿车为艺术摆设的陈列方式，设计师赋予了陈列室一个全新形象：以玻璃和金属为主要物料，将BMW最得心应手的三维设计艺术充分展示出来。陈列室的空间宽广，楼底高，即使加设了夹层，依然感觉宽阔。一条长直楼梯通往上层的销售处，顾客可在此俯瞰轿车的整体外观。楼梯的设计比例虽然轻盈，却成为陈列室里最瞩目的焦点，吸引着任何进入陈列室的客人目光。刻意显露的钢制天花梁底，其中一些部分安装了天花镶板，吊灯架亦按照建筑物的几何线条排列。轿车的陈列方式亦反映出这种流线形态，正好迎合德国人所强调的理性、准绳美学法则。

Detail

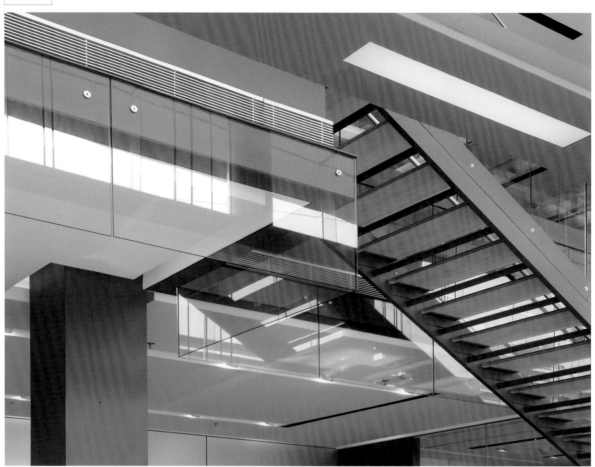

2003 | BMW showroom

Semir
chain store
森马服饰连锁店

China
中国

Completion Year : 2003 - 2007
Design : Alex Choi, Yan Yik Lun
Photographer : AlEXCHOI design & Partners

2003-2007 | Semir chain store

As one of the leading casual wear manufacturers in China, Semir has branch stores throughout the country. In the past, other than for the Semir name, these had been designed in a variety of styles depending on their location, but the company was keen to impose a more consistent look to help build a clearer brand identity for all its stores. The designers came up with a new system of co-ordinated display modules that would serve as prototypes for all outlets. The key was finding a way of marrying a high degree of flexibility with simple detailing, to create a cost-effective system that was easy to prefabricate, transport and install, while at the same time being versatile enough to cater for any number of different retail environments. With so many stores in many locations, there was little point in specifying floor and ceiling finishes except flagship stores, other than to suggest they should be as neutral as possible. The designers chose instead to concentrate on the three essential elements common to all outlets: a racking and shelving system to line the walls, a related system that could be applied to columns of different sizes, and a family of freestanding display islands. To keep everything simple, they developed a system based around rectangular metal frames that can easily be manufactured to size and transported in flat-pack form. A signature horizontal grill clothes rack was also developed to serve as a readily identifiable element unique to the brand.

In the company's larger stand-alone stores, the regular display racks and island units are combined with mannequins and special promotional units to create a more free-flowing space. Where special elements are required, they are designed to match the standard display units, so as the stair handrails, are in steel and have clear links to the metal frames used elsewhere to support wall racks and island units. Other than certain wall racks, few items are assigned fixed positions. Instead, they can easily be rearranged to create a sense of dynamism while maximising display space. The merchandise itself is then displayed in a seemingly casual and spontaneous manner that allows customers to make discoveries for themselves.

作为中国其中一间主要休闲制造商，森马的分店遍布国内，然而除了森马的招牌外，不同地点的店铺风格纷陈，欠缺一致性，故集团有意将所有分店的设计统一起来，以打造更鲜明的品牌形象。设计师替品牌设计一个全新的陈列模块，所有分店的成功关键在于设计必须简单而且拥有高度灵活性，符合经济原则，同时容易预制、运送及装嵌，以配合多样化的零售环境。由于分店的数目众多，除了旗舰店外基本上无法指定各分店所采用的地板及天花物料，故唯有将背景尽量保持中性，反而着重零售空间内必备的三维元素：靠墙的挂架及层架系统、可应用于不同体积的柱子的货架，以及一系列的独立式货岛。为简化设计，设计师构思了一个以长方形金属框为单元的陈列系统，既容易制造和装嵌，又可灵活调校陈列架的尺寸，更可平放以方便运送。设计师亦制作了一个条子屏风陈列架，令人一眼便能辨识。在较大型的专卖店内，除了标准的陈列货架及货岛外，更加入了模特儿公仔及其他宣传装置，令空间更自由流动。至于其他设计细节均为度身订造，以配合整体的陈列效果，例如楼梯的钢制扶手便呼应了靠墙货架及货岛的金属框。除却一些靠墙货架外，店内只有极少数的固定装置，其余大部分多为活动式设计，以便善用有限的陈列空间，营造出充满活力的购物氛围。货品亦采用了较随心即兴的陈列方式，鼓励顾客四处寻觅心头好。

2003-2007　Semir chain store

2003-2007 | Semir chain store

2003-2007 | Semir chain store

Fixture system

2003-2007 Semir chain store

K-Boxing men's fashion concept store

劲霸男士服饰概念店

Xiamen, China
中国厦门

Area : 643.82m²
Completion Year : 2009
Design : Alex Choi, Yan Yik Lun
Photographer : ALEXCHOI design & Partners
Brand Consultant : Shift

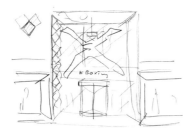

This grand looking store in Xiamen is the concept store of K-Boxing. The brand's positioning is mainly on mid to high-end men's fashion, especially smart casual wear. In order to enhance K-Boxing to a higher level, the logo has been renewed and has became a new image of the brand. The masculine logo was designed by a world-class advertising agency, and this masculine logo is extracted as a key element of the shop design. To promote the brand's new image, the new logo was furnished everywhere in the store; the door handle, the chandelier, the fixtures and display racks etc, anywhere you can think of. The brand's masculine logo was abstractly transformed into the artistic angular line, and skillfully applied through the store in and out. A systematic zoning design was practice in the store, the glamorous light boxes partition different sections and collections. These light boxes are not only pleasing to the eye but also practicable. In retail shop designs are not only to maintain the look and feel, but also the practicability. With the emphasis on quality of the brand, the design of the shop emphasizes details. The feature chandelier is one of the highlights, it is eye-catching and puts in the deluxe ambiance to the shop. Fine leather and stainless steel are used generally in the design of furniture and racks. These materials look cool, shining and elegant, and perfect for the expression of charisma; the fixtures are garnished with bevel edge pattern which emphasizing the delicate details.

这家位于厦门的大商店是劲霸男装的概念店。该品牌的定位主要是中高端时尚男装，特别是休闲服。为了将劲霸男装提高到一个更高的水平，标志已经修改，并成了一个全新的品牌形象。这个男性标志是由一个世界一流的广告公司设计的，它被提炼作为商店设计的一个关键元素。为了促进品牌的新形象，新标志装饰在店内各处：门的把手、吊灯、装置和显示架等任何你能想到的地方。该品牌的男性标志抽象地转化为艺术角线，巧妙地应用在商店内外。一个系统分区设计在商店被实践，迷人的灯箱隔开不同的部分和款式，这些灯箱不仅悦目而且实用。零售商店的设计不仅维持外观和感觉，而且是实用的。服装设计强调该品牌的质量，而商店的设计强调细节。这个特色吊灯是一大亮点，它是引人注目的，并将豪华氛围植入店中。好的皮革和不锈钢主要用于家具和机架的设计。这些材料看起来很酷、闪亮而优雅，完美地呈现着魅力；固定装置装饰着斜边模式，强调精致细节。

2009 | K-Boxing men's fashion concept store

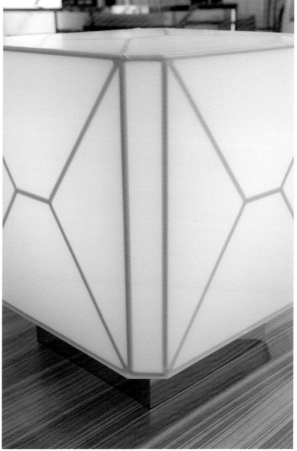

2009 | K-Boxing men's fashion concept store

Xplus shopping hub

Xplus 购物中心

Harbour city, Hong Kong, China
中国香港海港城

Area : 1,021.93m²
Completion Year : 2007
Design : Alex Choi
Photographer : AlEXCHOI design & Partners

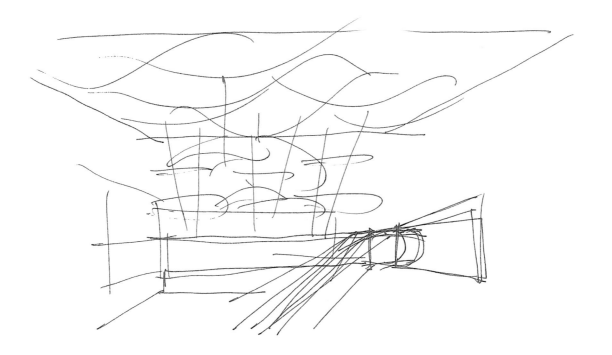

Xplus is a mini department store carrying a multitude of brands. The challenge was to create a clear brand identity for Xplus and the shops as a whole, while giving the individual products enough rooms to express their own special characters. Ocean Terminal, Tsimshatsui, had been selected as the site for the first store, which was generous in size but compromised by existing building services fixed to the soffit. In discussing a suitable concept, the phrase "everything under one roof" was raised up, which prompted the idea that a distinctive ceiling might be the key feature. It would provide the space with a single identity and conceal the services, while leaving the space below free for individual expression. An open waffle form was chosen to allow the services to remain hidden, but fully operational. The designers decided to "mould" the ceiling into a series of free-flowing undulations, resembling the gentle swell of ocean waves rolling across the entire space and creating a dynamic rhythm for the area below. Picking up on this oceanic theme, the display stands then became freestanding islands, softly organic in shape and designed as all-in-one units, incorporating storage and spotlights as well as a large counter-top. Square downlights inserted in the waffle ceiling provide lighting between the island units, which are supported on raking poles. At the edge of the space, the organic forms revert to a more rectilinear geometry, including the open grill that forms the shopfront, though here the undulating ceiling pushes out to create a sense of a space that is never static.

Xplus 为一家售卖不同品牌商品的小型百货公司，故项目的挑战在于如何为 Xplus 及其店铺构思一个清晰的整体形象，同时让个别产品及品牌展示各自独特的个性。客户选择了尖沙咀海运大厦作为首个店铺的所在地，空间虽然宽敞，然而天花却被大量楼宇喉管和电路所覆盖。在洽谈设计方向的过程中，客户特别提及"同一屋檐下"这个概念，灵机一动之下，设计师遂构思出一个特色的窝蜂形天花，不但将楼宇装备完全隐藏起来，也给予空间一个突出的形象，同时无碍地面空间的安排，容许各个品牌各自发挥陈列产品的创意。由于店铺相当大，我们将天花设计成恰如海洋上此起彼落的波涛，覆盖着整个空间，营造出起伏有致的韵律。为了贯彻海洋的主题，陈列架采用了独立货岛的方式，由数条柱杆支撑着，外形曲线流畅，且集储物柜、射灯照明及陈列台面多功能于一体。窝蜂形天花安装了方形下照射灯，以弥补陈列架射灯光线的不足。越接近店铺的边缘，有机的曲线形态逐渐被直线几何所替代，店面入口的条子屏风便是最佳例子，而起伏不定的天花更越过边界，往店外延伸，营造出跃动不息的空间感。

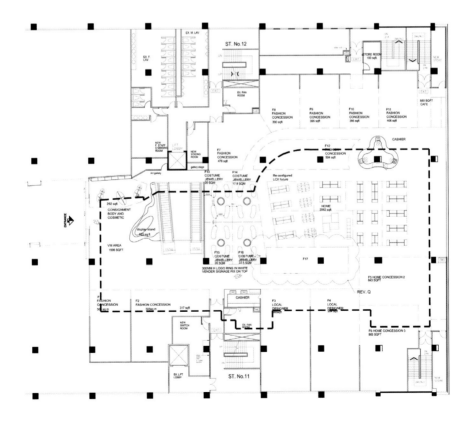

Xplus shopping hub | NO.34 - 35

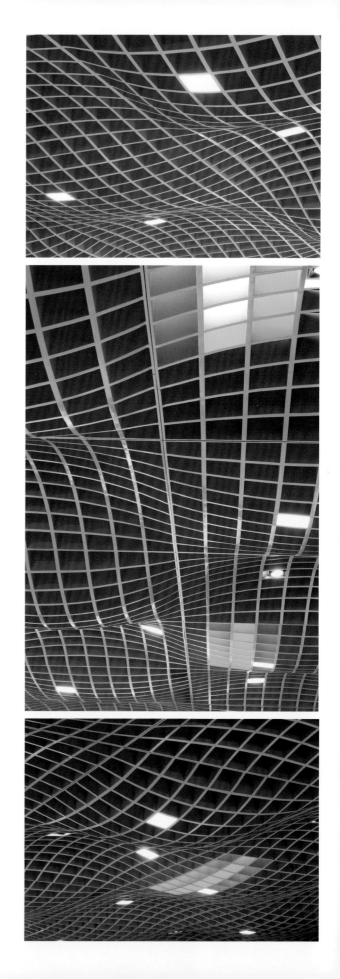

OFFICE
办公室

KML Engineering Ltd. office
高明科技工程有限公司办公室

Global Star Entertainment & Technology office
Global Star Entertainment & Technology 办公室

Studios of Centro Digital Pictures
先涛数码影画制作有限公司工作室

C.F.L. Enterprise Limited office
赛辉洋行有限公司办公室

Artapower International Group Ltd. office
艺达堡集团有限公司办公室

Pearltower Garments & Toys Co. Ltd. office
宝台制衣玩具有限公司办公室

Artapower International Group Ltd. headquarter
艺达堡集团有限公司总部办公室

Jeanswest International (H.K.) Ltd. - China head office
真维斯国际(香港)有限公司 – 中国总部办公室

KML Engineering Ltd. office

高明科技工程有限公司办公室

Hong Kong, China
中国香港

Area : 726.50m²
Completion Year : 2000
Design : Alex Choi
Photographer : ALEXCHOI design & Partners

The strategy behind the refurbishment of this office in an industrial building stems from the nature of the company's business in high quality metalwork. Drawing on this tradition elaborate metal and glass details abound, conveying futuristic, hard-edged impression that befits the client's engineering background. The design also takes advantage of the generous 5m ceiling height to accommodate a suspended mezzanine floor. This houses the meeting rooms, allows the main office floor to be given over exclusively to informal workstations. Matching the central mezzanine, a "floating" glass cube is installed to one side of the main space to house the director's office. This changes the dynamic of the space around it, while affording the office abundant light and stunning views of the river outside. From the acoustic ceiling panels that conceal the lighting and air-conditioning services to the elements that make up the suspended mezzanine floor, the design of every detail is the result of a collaborative dialogue between designer and client that sought to push the boundaries of metal engineering to new heights.

这一间办公室位于一工厂大厦内，公司从事高质素金属制品业务，办公室的整修亦以此为出发点，细致的金属及玻璃细节随处可见，塑造出充满未来感的高科技形象，亦正好切合公司的工程背景。为了善用5m高的楼底，空间特别加设了夹层，用作会议用途，令整个主楼层可作为员工的办公室空间。为了配合夹层的安排，董事总经理的房间设计成仿如玻璃立方体，"悬浮"于办公室的其中一边，顿时改变了空间的气场，同时令房间可坐拥充沛的阳光和毗邻河道的景观。由覆盖照明电线及冷气管的隔音天花屏板，以至支撑夹楼的金属结构，所有细节都由设计师及客户共同研发，发掘金属工程在设计中的种种崭新可能。

2000 | KML Engineering Ltd. office

Global Star Entertainment & Technology office

Global Star Entertainment & Technology 办公室

Hong Kong, China
中国香港

Area : 906.55m²
Completion Year : 2000
Design : Alex Choi
Photographer : ALEXCHOI design & Partners

2000 | Global Star Entertainment & Technology office

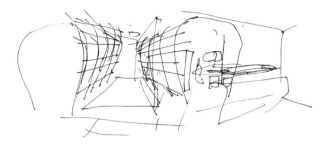

This new office for a cutting-edge design company had to fulfill two programmatic needs: to accommodate a large number of staff in a small space, and to serve as a natural advertisement for the company. The interior is organized around a series of large arches enveloped by semi-transparent polycarbonate to maintain a degree of privacy. These instantly steer away from the rigidity of a conventional office and distinguish the company's funky, playful attitude. Interspersed throughout the space are communal workstations, where hierarchy has been deliberately deconstructed to encourage creativity and interactivity. That sense of spontaneous creativity is further enhanced by the white epoxy floor, which reflects both the bluish glow cast by the fluorescent lights and the company's cool, youthful energy.

这间构思前卫的办公室，设计上需要顾及两个实际需要：在狭小的空间里容纳所有员工，同时成为公司的一个"招牌"。室内设有一连串以半透明聚碳酸酯制成的拱形结构，打破了一般办公室的生硬布局，亦替公司塑造出破格、玩味的形象。至于员工共享的工作台间则分散其中，打破了固有的阶级观念，促进双方的创意及互动交流。白色环氧树脂地板进一步加强这种随意即兴的感觉，同时倒映出光管投射出的蓝光，为公司注满年轻的正能量。

2000 | Global Star Entertainment & Technology office

Studios of Centro Digital Pictures

先涛数码影画制作有限公司工作室

Hong Kong, China
中国香港

Area : 308.44 m²
Completion Year: 2000
Design : Alex Choi
Photographer : ALEXCHOI design & Partners

Centro is a Hong Kong-based animation company providing creative visual effects to the movie and reflected in the design of their early Causeway Bay office, however, which deliberately avoided futuristic style cliches and focused instead on creating a more homely aesthetic - a comfortable workplace for employees who often have to work very long hours. Refurbished from old tong lau, the traditional mezzanine was retained as a meeting room, while the staff studios were located to the rear on the ground floor. A staircase of natural wood and crisp metalwork connected both levels, establishing a palette of materials that was used throughout, including in the informal "meet & greet" area beyond the reception that was included to promote greater interaction - an ethos that the company strongly advocates.

以香港为基地的先涛，专门为电影及互动媒体行业提供创新的数码视觉特技制作，他们位于铜锣湾的首个办公室，却丝毫不见老掉牙的未来主义幌子，设计反而着重营造一种舒适的家居气氛，特别是为经常需要长时间工作的员工而设。办公室由一幢唐楼改建而成，典型的阁楼层被保留下来，现在变成会议室，而员工的工作室则位于地面层的末端。由木材和金属组成的轻巧楼梯贯穿两层空间，同时主导了室内其他地方的物料选择，包括毗连接待处的交流区，有助于加强员工之间的互动沟通，亦是公司一直推崇的企业宗旨。

Studios of Centro Digital Pictures

2000 | Studios of Centro Digital Pictures

2000 | Studios of Centro Digital Pictures

C.F.L. Enterprise Limited office

赛辉洋行有限公司办公室

Shenzhen, China
中国深圳

Area : 2,683.04m²
Completion Year : 2003
Design : Alex Choi
Photographer : ALEXCHOI design & Partners

2003 | C.F.L. Enterprise Limited office

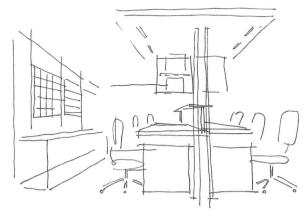

This office for a clothing manufacturer in Hong Kong plays with the concept of using furniture and material finishes as the main anchoring design features. The office's footprint is long and deep, extending nearly 30.48m but with an extremely low ceiling. To help break up this imposing volume, individual, purpose-designed workstations line the centre of the space, flanked on both sides by continuous corridors and semi- or fully-enclosed booths and rooms for senior management and private meetings. Suspended wooden cabinets and occasional timber panels add warmth and textural contrast, while a patchwork of different wood veneers add interest and help identify individual workstations. Above the corridors and between some of the central workstations, the ceiling is lined with easily demountable, white plastic tubes that not only echo the linearity of the office in an instantly memorable way, but also conceal the mechanical services above. Over the workstations themselves the tubes give way to solid panels, brightly illuminated by concealed uplights between the hanging cabinets to counteract the low ceiling height and prevent glare. Along the central space, coloured glass panels are employed to further break up the length of the office and identify particular area. Blue denoted the purchasing department, for example, while red and orange represent the administrative and delivery departments respectively, helping to create and individual identity for each area while promoting a sense of interactivity and exchange.

这间香港成衣制造商的办公室，利用家具和物料作为主导的设计概念。办公室的楼底不高，楼面既长且深，总长度差不多有30.48m。为了消除空间的狭隘感，度身设计的独立工作台被安排在中央位置，两边是走廊，继而是半开放式或密封式的经理室及会议室。各工作台均配置木吊柜，加上间或出现的木墙板，加强了材质的对比；不同薄木片形成的拼贴效果，亦方便辨识不同部门的所在位置。走廊上方，以及部分工作台之间的天花，整齐排列着可拆除的白色塑胶管，不但回应了空间的直线布局，同时亦将电线和喉管隐藏起来。工作台上的吊柜装置了隐藏式上照灯，将灯光投射到木天花上，令楼底感觉较高，更有效地改善眩光问题。办公室内各部门以不同颜色的玻璃屏风标示，例如蓝色为采购部，红色为行政管理部，而橙色为送货部，打破了空间的单调线条，亦加强了部门之间的互动和交流。

2003 | C.F.L. Enterprise Limited office

Work station

2003 | C.F.L. Enterprise Limited office

Artapower International Group Ltd. office

艺达堡集团有限公司办公室

Hong Kong, China
中国香港

Area : 1,338m²
Completion Year : 2005
Design : Alex Choi
Photographer : ALEXCHOI design & Partners

2003 | Artapower International Group Ltd. office

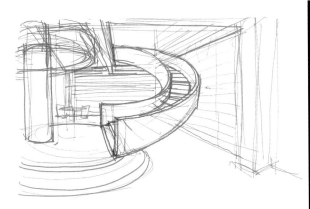

Artapower is a toy manufacturer who has headquarter office and production lines in Jiangmen, another production factory in Shenzhen and design development office in Hong Kong. So this Hong Kong office is designed as a "meeting & greeting" place between their toy designers and the U.S. clients. Open, relax and casual office space is the goal to create. Reception, general workstations and meeting rooms are located on the upper floor while lower floor accommodated showrooms, open kitchen and brain-storming room. Two floors are connected with a dramatically designed light weight staircase in curve profile. It associates with the round "Totem" to become the landmark of the office. The building is a 30 years old factory building with high ceiling. Light bulbs in pendant style with a wire "Loop" are used as general light source which help to exaggerate the high ceiling as well and also an unconventional treatment on office lighting. Bleached fibre board is the material for the general partition which has a sense of casual, vivid colour and dinky graphics are used to enrich the space and convey a sense of "Toy".

艺达堡是一家玩具制造商，在江门设有总部办公室和生产线，在深圳设有分工厂，并在香港设有设计开发办公室。因此，其香港办公室旨在打造为玩具设计者与美国客户之间进行"会面与问候"的场所。目标定位为开放、放松、休闲的办公空间。接待室、公共工作站和会议室位于上层，下层有陈列室、开放式厨房和头脑风暴区。两层之间以曲线轮廓的轻质楼梯相连。曲线楼梯与圆形的标识相呼应，成为办公室的亮点。楼层是30年前的旧工厂，有着高高的天花板。悬吊式灯泡提供主要光源，使得天花更显其高，也使得办公室照明与众不同。漂白纤维板用于充当隔板，增加休闲和动态之感。极小的图形图案用于丰富空间，并营造"玩具"感。

2003 | Artapower International Group Ltd. office

Pearltower Garments & Toys Co. Ltd. office

宝台制衣玩具有限公司办公室

Shenzhen, China
中国深圳

Area : 3,000m²
Completion Year : 2005
Design : Alex Choi
Photographer : ALEXCHOI design & Partners

2005 | Pearltower Garments & Toys Co. Ltd. office

In contrast to the Artapower offices, the four-storey headquarters for sister company Pearltower, in Shenzhen, is filled with colour. Red, yellow, green, blue and orange are used liberally throughout the building as walls, screens, doors, and even furniture, reflecting the company's business as a ferment and toy manufacturer while also serving as an effective means of orientation and wayfinding. However, the principle of respecting the building's structure has not been forgotten. Beams are deliberately exposed and highlighted by fluorescent lights concealed in their undersides, while stairs and beams criss-cross the central atrium leading to the factory above, adding a sense of movement to this dynamic space.

项目位于深圳宝台制衣玩具有限公司的4层高总部大楼，跟其姊妹公司艺达堡的办公室截然不同，红、黄、绿、蓝和橙等颜色充斥空间每一角落，包括墙壁、屏风、大门甚至家具，正好符合公司作为一家成衣及玩具制造商所需要的企业形象，同时协助员工辨识方位。然而以大厦的建筑结构作为设计主体的原则依旧不变，横梁没加修饰地暴露出来，梁底亦隐藏了光管，加上通往上层厂房的楼梯左穿右插于中庭之中，令空间更添动感。

Atrium

2005 Pearltower Garments & Toys Co. Ltd. office

2005 | Pearltower Garments & Toys Co. Ltd. office

Artapower International Group Ltd. headquarter

艺达堡集团有限公司总部办公室

Jiangmen, China
中国江门

Area : 3,275.48m²
Completion Year : 2005
Design : Alex Choi
Photographer : Ken Choi

2005 | Artapower International Group Ltd. headquarter

The design of the Artapower Headquarters in Jiangmen hinges upon a single word — exposure. Within this three-storey industrial building, all secondary finishes and coverings have been stripped away, leaving behind only the exposed structure and its waffle floor-slabs, that, to the designers at least, seemed very beautiful. The walls and floors are finished in a simple palette of white and grey, against which the waffle slabs, columns and staircases become the main elements in defining the space, as well as dramatic features in their own right. Circular light pendants are chosen to contrast with the rectangular geometry of the waffle slab, while the austerity of the industrial environment has been broken with occasional splashes of colour in the form of playful pieces of furniture and painted panels that adorn some of the doors and walls.

艺达堡江门总部的设计基于一个关键词——展露。楼高3层的工厂大厦，所有内部粉饰都被拆掉，剩余必要的承重结构及窝蜂楼板，这些元素对设计师而言均充满建筑美感。墙壁和地板髹以简单的灰、白色调，令窝蜂楼板、柱子及楼梯成为空间最主要及唯一的装饰元素。圆形天花灯与窝蜂楼板的几何直线形成有趣的对比，而颜色鲜艳的家具、大门及墙壁则为空间注入活力，打破严肃的工业氛围。

2005 | Artapower International Group Ltd. headquarter

Jeanswest International (H.K.) Ltd. - China head office

真维斯国际(香港)有限公司 —— 中国总部办公室

Huizhou, China
中国惠州

Area : 1,858.06m²
Completion Year : 2005
Design : Alex Choi
Photographer : ALEXCHOI design & Partners

2005 | Jeanswest International (H.K.) Ltd. - China head office

Glorious Sun Enterprises is the company behind the clothing brand Jeanswest, and its two-storey, 1,858.06m² headquarters in Huizhou houses 100 staff. In order to encourage an open exchange of ideas, the project required a design that was practical and creative at the same time. A standard, open-plan, "cubicle" layout was adopted for its sufficient use of space, but any sense of ordinariness was offset by creative and playful detailing. The expansive high ceiling is accentuated with diffused light troughs that follow the flow of circulation routes below, while mushroom-shaped structural columns – which further strengthen the sense of height – act as a counterpoint to the rectilinear layout. The corporate colour blue abounds and serves different purposes. Blue glass demarcates the managers' offices and offers greater privacy, while lighthearted blue signs promote a relaxed and friendly environment that befits the company's image.

Glorious Sun Enterprises 为时装品牌真维斯的母公司，其位于惠州两层高总部大楼内，占地面积为 1,858.06m²，员工人数达一百名。为了鼓励员工之间的创作交流，故项目要求设计必须兼备实用性及创意思维。办公室虽然采用了标准的开放式工作间间隔，然而细节处理却处处流露出不平凡的心思，如天花的隐藏式灯槽与地面的走廊位置互相呼应，结构柱被粉饰成蘑菇状，不但令楼底显得更高，亦与方正的平面布局形成鲜明对比。室内随处可见集团的企业颜色——蓝色，并且各具功能：经理房间缀以蓝色玻璃，增加私密感；为配合公司的形象，路标同样以蓝色为主调，营造出极具活力的工作氛围。

2005 | Jeanswest International (H.K.) Ltd. - China head office

EXHIBITION
展览厅

Young Achievers' Gallery
香港教育局荟萃馆

CLP Energy Efficiency Exhibition Centre
中华电力能源效益展览中心

CLP Smart Grid Experience Centre
中华电力智能电网体验馆

Young Achievers' Gallery

香港教育局荟萃馆

Hong Kong, China
中国香港

Area : 2,675.60m²
Completion Year : 2003
Leading Consultant : Centro Digital Pictures Ltd.
Interior, Lighting & Exhibition Design : ALEXCHOI design & Partners
Photographer : ALEXCHOI design & Partners

Driven by the desire to create an "outside-the-box" exhibition space, the designers actively sought innovative ways to showcase award-winning performances in a variety of disciplines, displaying each in the most exciting way possible. Established in 2007, the gallery aims to inspire young visitors to strive for excellence and achieve full potential in their particular field, reflected by the seven "domains" covering the Arts, Design, Language and Humanities, Mathematics, Science and Technology, Sports and Thinking. In the Arts domain, for example, a piano with in-built speakers is suspended from the ceiling, while other musical instruments are treated as sculptures within frames to create an appropriately theatrical air. Throughout the exhibition various interactive installations have been devised to engage, excite and inform. Display boards have been replaced by touch-sensitive LCD monitors on which visitors can activate, at their own pace, moving images and videos describing each award-winner's success story. The whole experience culminates in a mini-theatre with a 180-degree panoramic screen, guaranteed to leave a lasting impression on all visitors.

为了摆脱传统的展览空间模式，我们致力于破旧立新，将学生在不同领域的卓越表现和成就，以最新的方式展示出来。荟萃馆展览中心创立于2007年，目的是鼓励年轻的参观者发掘独特的个人潜质，以及培育他们锲而不舍，精益求精的态度。展览内容涵盖七大范畴，包括艺术、设计、语言及人文学科、数学、科学及科技，以及体育及思维，例如在艺术的区域，一个内置了扬声器的钢琴悬吊于天花板，其他乐器则犹如雕塑般被裱框起来，为空间注满戏剧性的艺术气息。展览中心到处均设有互动装置，吸引参观者主动观赏、查阅，带给他们不同的感官刺激，触控式LCD荧幕代替了传统的展板，参观者可按各自的参观步伐，启动装置，观赏阐述各位得奖学生的有关图片和录像。参观者最后会来到一个小型剧院，剧院里设置了180°全景屏幕，令所有参观者对屏幕上播放的影片留下更深刻的印象。

2003 | Young Achievers' Gallery

2003 | Young Achievers' Gallery

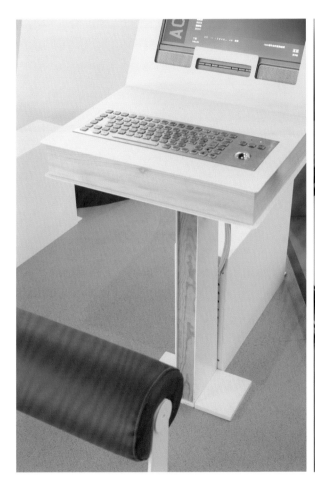

CLP Energy Efficiency Exhibition Centre

中华电力能源效益展览中心

Hong Kong, China
中国香港

Area : 260.12m²
Completion Year : 2009
Design : Alex Choi
Photographer: Bobby Wu

2009 | CLP Energy Efficiency Exhibition Centre

"Bubbles" is the key design element which is derived from the bubble diagram at the very early stage of design development. The client's idea is very demanding since they asked for putting 7 zones in a 260.12m^2 area while each zone can accommodate a tour with a maximum of 30 persons. Overlapping of main bubble spaces reduces the sense of congestion. The curve route naturally created prevents the visitors from viewing all the zones at one glance. They have to explore the zones one by one, and it is hard for them to recognize the size of the whole space. In accordance with the "bubbled" layout, bubble elements are introduced to the custom-made ceiling tiles which help to wrap up different zones and also conceal the light, sprinkler head and air grille which usually ruin the design. Stepping on the entrance, walk through the aisle which is coated with the green and blue patterned wall and feel the breeze of this green ambiance. Behind the black curtain, immediately catch your sight is the impressive "bubbles" ceiling and visitors are so ready for a high technological adventure. Systematic arrangement strategy is applied in this centre, every particular zone of the showrooms is clearly parted. Artistic Chinese characters show the theme of each zone and unique visual effects help to standout different parts of the exhibition. Natural clockwise flow is costumed made for visitors to explore the showrooms one by one without missing a thing and this magical spatiality tricks visitors' senses into believing they are in a huge exhibition centre instead of pacing in the 260.12m^2 area.

作为能源效益展览中心，设计师选择以"泡泡"作为其关键的设计元素，这个灵感源于早期设计的泡泡图形。按客户的要求，在260.12m^2 的区域内必须划分出7个展览区域，而且每区需容纳不少于30名参观者。为符合公司要求，并兼顾实用性能和美观的整体效果，设计师利用几个主要的泡泡型空间的相互重叠来减少拥挤感。自然而成的曲线过道遮蔽了参观者的部分视线，不让他们一下子看到所有区域，而必须逐步探究。泡泡图案造型天花既迎合泡泡型的布局，亦可以用来隐藏灯具等容易影响整体效果的装置。在没有泡泡造型天花的展区内，萤光绿色刷漆管道成为一道亮丽的风景线，和整个展厅的蓝绿色主色调相互呼应。在面积不大的展区内，完成了既符合客户要求，又实用美观的空间，给来访者留下一段愉快的参观体验。

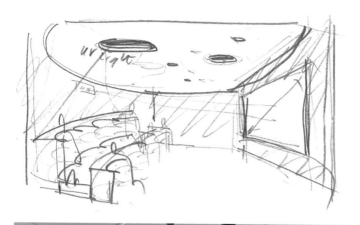

2009 | CLP Energy Efficiency Exhibition Centre

CLP Smart Grid Experience Centre

中华电力智能电网体验馆

Hong Kong, China
中国香港

Area : 213.6m²
Completion Year : 2011
Design : Alex Choi
Photographer : Bobby Wu

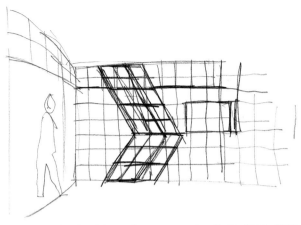

As one of the biggest company in Hong Kong that supplies half of the city high quality and safe electricity to make people in town live better lives, this technology driven company also takes the mission on delivering the right message to the public. When people talk about education to the public, the first thing that comes to mind may possibly be: exhibition. Beside supply electricity, there are many back up technology and support services that the company would like to introduce to professional bodies. So much technology information may sometimes sound boring, so the designers' mission is translating those technical driven information to some interesting information that can be easily digested but also keep them informative. Delicate leading paths are applied to all CLP exhibition centres, visitors will be escorted by technicians and they can also explore through over the space. Interaction always holds the spirit of the whole exhibition, visitors will have so much immediacy and warmth while the display content can interact with them, and leaving deep impression to the visitors is the core of a successful exhibition. But after all, the designers did a great job to their hidden mission:implanting the professionalism corporate image in the exhibition area. Just like air, you cannot resist to breath and you take in willingly.

作为香港最大的电力供应公司之一，为差不多半个城市的市民提供高质量的安全电力服务，让人们过更好的生活。作为一间科技公司，也承担着把正确的信息传递给公众的任务。当谈及公共教育，人们的脑子里首先想到的可能是：展览。除了供应电力，公司还有很多技术性支持服务想介绍给专业机构。这么多技术信息有时候听起来会感觉无聊，所以设计师的任务是将这些以技术为主的信息转化成既容易消化又可以增长知识的有趣信息。所有的中华电力展览中心都应用了精致的主通道，有时技术人员会引领访客，又或是让访客自由探索空间。互动是整个展览馆的灵魂，当显示内容与他们产生共鸣时，访客会感到分外亲切，而令访客留下深刻印象正是这个成功的展览馆的核心价值及成功之处。设计师们的隐藏任务亦表现出色：将专业的企业形象植入展览区。就像空气一样，你无法抗拒呼吸，甘愿接收。

2011 | CLP Smart Grid Experience Centre

RESIDENTIAL
住宅

A Private Villa in Santa Monica, L.A.
洛杉矶圣塔莫尼卡私人别墅

Bachelor's flat in Kowloon Tong
九龙塘住宅项目

A Private Villa in Santa Monica, L.A.

洛杉矶圣塔莫尼卡私人别墅

Los Angeles, USA
美国洛杉矶

Area : 729.66m²
Completion Year : 2005
Design : Alex Choi
Photographer : Alex Choi, Dana Hoff

2005 | A Private Villa in Santa Monica, L.A.

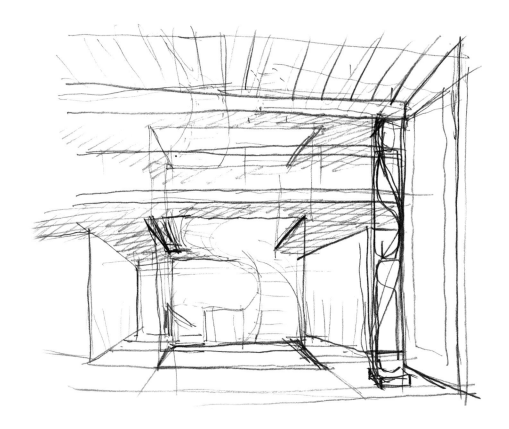

Light and ventilation are the key concepts behind this house for a family of four in Los Angeles. Designed in association with the Californian architect, Bridgwater Consulting Group, the three-storey building takes full advantage of the surrounding landscape and abundant sunlight, with windows carefully oriented to establish a dialogue between the indoor and outdoor spaces and the hillside views beyond. They also flood the interior with light and, when open, allow the house to breathe in the cooling sea breezes. All the main living spaces revolve around a spiral staircase, a powerful sculptural centerpiece whose organic form stands in stark contrast to the otherwise rectilinear layout. Topped by a generous skylight, the stair also acts as a lightwell, allowing daylight to penetrate deep into the building while facilitating natural ventilation throughout. Stone floor tiles, timber clad ceilings and white stucco walls comple the sleek modernity of the house, creating a neutral backdrop for the play of light and shadow, highlights with occasional accents of bright colour.

保持光线充沛和空气流通，是这个位于洛杉矶的4口之家的主要设计考虑因素。房子楼高3层，由加州Bridgwater Consulting Group设计，为了引入房子周围的开敞景观及和煦阳光，窗户的位置均经过精心设计，令室内与室外，以至远处的连绵山峦，都能建立互动性，同时也让室内洒满阳光，其中更有海风吹拂。主起居空间围绕着一条旋转楼梯，楼梯恰如雕塑的弧线造型，跟房子的正方形态形成鲜明对比。楼梯顶端设有一个天窗，令楼梯仿如采光井，光线可透过天窗渗进房子的任何角落，同时有助于自然通风。石地砖、木砌天花及白色石纹墙壁，延续了房子的流畅现代氛围，为光与影提供了一个恰当背景，还偶尔加进点点鲜艳色彩作为点缀。

A Private Villa in Santa Monica, L.A.

2005 | A Private Villa in Santa Monica, L.A.

2005 | A Private Villa in Santa Monica, L.A.

2005 | A Private Villa in Santa Monica, L.A.

Work in progress

2005 | A Private Villa in Santa Monica, L.A.

Bachelor's flat in Kowloon Tong

九龙塘住宅项目

Hong Kong
香港

Area : 112 m²
Completion Year : 2006
Design : Alex Choi
Photographer : Alex Choi

This 112m² apartment for a young couple was designed to maximize the use of space, while reflecting the owner's particular lifestyle needs. This was achieved by removing most of the non-structural walls to create a blank canvas in which the various functional areas could be freely located. The main living/dining room, therefore, became a single fluid space that opened directly onto an existing full-width balcony with lush greenery beyond. Similarly, conventional doors were replaced with large sliding panels of sandblasted glass that could be easily pushed aside to reveal the master and guest bedrooms and include them as part of the main living space. The primary finishes were deliberately plain and simple, with self-levelling concrete for the floor and whitewashed walls and ceilings, creating a neutral backdrop against which the owners' collection of art and furniture took centre stage, shining through the space.

这间面积约112m²的房子，设计主题除了要善用空间外，同时也反映了屋主夫妇二人的独特生活习惯。方法是将室内大部分的非承重墙拆去，营造出素净的背景，让不同的功能空间能够随意配置，例如，客厅便成为一个流动的空间，一直延伸至露台以至屋外的茂密绿荫。为了贯彻这个氛围，设计师摒弃了传统的推拉门，改以磨砂玻璃趟门将主卧及客房围合起来，同时可与主要起居空间连成一体。物料的选择简单朴实，地板为自动流平混凝土，墙壁及天花同为白色，在一室素白下，令屋主收藏的艺术品和家具更显突出。

2006 | Bachelor's flat in Kowloon Tong

Acknowledgements

We would like to thank all the designers and companies who made significant contributions to the compilation of this book. Without them, this project would not have been possible. We would also like to thank many others whose names did not appear on the credits, but made specific input and support for the project from beginning to end.

Future Editions

If you would like to contribute to the next edition of Artpower, please email us your details to: artpower@artpower.com.cn

ALEXCHOI design & Partners

collections
蔡明治设计精选作品集

Volume Two
Coherence
第二册 建构和谐

深圳市艺力文化发展有限公司 编

华南理工大学出版社
SOUTH CHINA UNIVERSITY OF TECHNOLOGY PRESS
·广州·

We used our pens as the space edit tools & built up the scenario structures one by one, it is about presenting its own personality & we translated the message into interior.

SINCE 1997

COHERENCE

————————————————————— SINCE 1997

图书在版编目（CIP）数据

蔡明治设计精选作品集 = ALEXCHOI design & partners collections.2，建构和谐 / 深圳市艺力文化发展有限公司编．— 广州：华南理工大学出版社，2014.4
ISBN 978-7-5623-4118-5

Ⅰ．①蔡… Ⅱ．①深… Ⅲ．①室内装饰设计－作品集－中国－现代 Ⅳ．① TU238

中国版本图书馆CIP数据核字（2014）第 059359 号

蔡明治设计精选作品集 ALEXCHOI design & Partners collections
第二册 建构和谐 Volume Two Coherence
深圳市艺力文化发展有限公司 编

出 版 人：	韩中伟
出版发行：	华南理工大学出版社
	（广州五山华南理工大学 17 号楼，邮编 510640）
	http://www.scutpress.com.cn　E-mail: scutc13@scut.edu.cn
	营销部电话：020-87113487　87111048（传真）
策划编辑：	赖淑华
责任编辑：	陈　昊　赖淑华
印 刷 者：	深圳市汇亿丰印刷包装有限公司
开　　本：	787mm×1092mm　1/16　**印张**：22.75
成品尺寸：	185mm × 260mm
版　　次：	2014 年 4 月第 1 版　2014 年 4 月第 1 次印刷
定　　价：	398.00 元（共 3 册）

版权所有　盗版必究　　印装差错　负责调换

CONTENTS

ENTERTAINMENT
娱乐

Luoyang Wanda International Cinema   2
洛阳万达国际影城

HOSPITALITY
餐饮休闲

Aveda + Xenses Lifestyle Spa & Salon   10
Aveda + Xenses 发廊及水疗中心

Aropa Mediterranean Restaurant   14
Aropa 地中海餐厅

Suzuki café chain   18
铃木珈琲馆连锁店

Caffè HABITŪ Chain   22
Caffè HABITŪ 连锁咖啡厅

Caffè HABITŪ — the table   30
Caffè HABITŪ 咖啡店——餐桌

RETAIL
零售

3 shop chain store   32
3 shop 移动电讯连锁店

Sole Town   38
Sole Town 鞋履配饰专卖店

EQ:IQ flagship store   42
EQ:IQ 服饰旗舰店

EQ:IQ chain store   48
EQ:IQ 服饰连锁店

Xenses cosmetic shop   64
Xenses 美妆品专卖店

GENIO LaMode men's fashion concept store   68
GENIO LaMode 男士服饰概念店

Marlboro Newstand   72
万宝路书报摊

Marlboro Counter   78
万宝路香烟柜台

K11 Design Store   82
K11 设计品味专卖店

OFFICE
办公室

Jeanswest International (H.K.) Ltd. Office   90
真维斯国际(香港)有限公司办公室

ALEXCHOI design & Partners office   94
蔡明治设计有限公司办公室

RESIDENTIAL
住宅

Apartment in Harbourside, Hong Kong   102
香港君临天下住宅项目

ENTERTAINMENT
娱乐

Luoyang Wanda International Cinema
洛阳万达国际影城

Luoyang Wanda International Cinema

洛阳万达国际影城

Luoyang, China
中国洛阳

Area : 3,530m²
Completion Year : 2010
Design : ACD X CNI
Lead Designer : Alex Choi - ALEXCHOI deaign & Partners
Photographer : Ken Choi, Guozeng Jiang

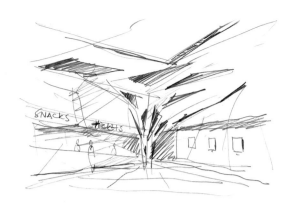

"Dream" is the design concept of the Luoyang Wanda International Cinema. The fifth generation film director in China - Yimou Zhang has once said that: "Movies are dreams". Movie itself is a dream, a fantasy, and it gives people endless imagination. The ultimate aim for this project is to create a wonderland which people can leave all the trouble and worry behind, fall into the movie and enjoy the imagination with the pleasure. Inspired by the "Hong gao liang" (also named "Red Sorghum"), which is Director Yimou Zhang's first movie. The movie was a great success; it had won many international awards and became a mile-stone of China movie in the world, so it is the most representative of China movie. Red is used as the key tone of the cinema, it not only matches the theme of "Red Sorghum", but also matches with the cooperate color of Wanda. The innovative Chinese style brings the new wave to Luoyang, people will immediately experience the surrealistic ambience once they set their feet into the cinema. The irregular geometry pattern decorates the ceiling which is transformed into the illustrative sorghum, plus the LED red light exaggerates the dimension of the sorghum ceiling. In order to stand out the geometry style, the wall finishing is also filled with crisscross linear art. Lighting becomes the partition between the lobby and theaters. The irregular geometry pattern ceiling finishing was use to connect the aisles and lobby; using the bright light in the lobby is to gather the audiences, walking on the aisles to the theaters, the darken light was the sign to the audience to prepared themselves for the movie. The shapes of the theatre entries were designed as unique, stylish geometric boxes so as to implement the whole geometric style. Well use of red and yellow for lighting and wall finishing is a perfect match of the Wanda International Cinema, it presents the absolute balance of funny and fancy but never fussy.

万达国际影城的洛阳项目的设计理念来自"梦"。中国著名第五代大导演张艺谋曾经说过："电影跟梦，其实是一样的"，电影本身就是一个个的梦想、幻想，而电影也赋予人们无限的想象。所以这项目的设计目的，是要设计一个让人可以远离烦嚣、忘却烦忧的空间，在这个写意之地愉快地欣赏精彩的电影，尽情地发挥想象空间。整个设计的灵感取材自第五代大导演 — 张艺谋第一次执导的电影"红高粱"。这部电影在国际上获奖无数，创造了中国电影的一个高峰，是中国电影在国际舞台上的一个重要的里程碑，所以这部电影最能够代表中国电影的发展。整体设计以红色作为影城的主调，除了配合电影的主题外，也提取了万达标志中的红色作为设计的元素。这个创新、破格的概念，时尚又充满中国风的超前设计，为洛阳带来一个新浪潮，使人们踏进影城就已经有超现实的体验，甚至可以说连欧美等国也难以媲美。以抽象化的高粱外形，设计成不规则的几何造型天花。大堂的几何造型天花在成本控制及扩展上有一定的困难。所以在未来的设计上，建议可以把几何的立体天花造型简化，让立体的以较为平面的方式表现，不但可以保留几何作为主题，按照不同地点的环境，以同一个设计方向，做出不同的效果；简化后的天花造型不但可以用较低的成本推广至第二及第三城市，更可让施工更为容易。大堂墙身纵横交错的凹槽造型饰面，形象突出之余又可以贯彻整个几何设计的风格。大堂的抽象化仿高粱的几何天花造型，加上红光来使立体感更为突出。大堂的LED招牌标示，使用LED灯比一般的光管出光效果更闪、更突出，在用色上也可以有更多变化。用上可变色的LED灯，即可以随不同的节日、时间，用不同的灯光颜色来营造特别的视觉效果和气氛。利用灯光把大堂和影厅分隔开；大堂以光亮的灯光把人们凝聚，影厅的走廊则用较暗的灯光，让人们可以在进入影厅以前酝酿情绪欣赏电影；大堂跟影厅以几何造型天花贯通两个不同的空间，使整个影城更统一。大堂幽幽的灯光，以较温和的手法营造出大堂和影厅在气氛上的对比。影厅走廊内用上标准化开放式天花设计，把原本的天花喷黑，再挂上标准化几何形特色吊挂天花，配合较暗的灯光营造出幽静的格调。使用这种标准化的天花造型设计，既切合潮流，施工亦很简单，对于在拓展业务时既可以减少施工时的繁复步骤，也更容易控制成本。标准化的标示牌分布平均，无论文字及数字都表达清晰。几何盒子造型的标准化影厅入口及天花挂吊厅号牌，外型超前、独特，既切合潮流，亦贯彻了整个几何风格；加上以万达标志上的红色作为主调，突出万达的形象。灯光及墙身用红色及黄色做主调，绝妙的颜色搭配使影厅的形象及气氛统一，亦免于给人太花哨的感觉。洛阳万达国际影城的完美标准化标示系统，绝对可以应用在未来每一个万达国际影城。

2010 | Luoyang Wanda International Cinema

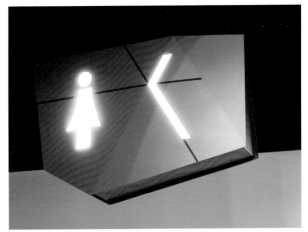

HOSPITALITY
餐饮休闲

Aveda + Xenses lifestyle spa & salon
Aveda + Xenses 发廊及水疗中心

Aropa Mediterranean restaurant
Aropa 地中海餐厅

Suzuki café chain
铃木珈琲馆连锁店

Caffè HABITŪ chain
Caffè HABITŪ 连锁咖啡厅

Caffè HABITŪ - the table
Caffè HABITŪ 咖啡店—餐桌

Aveda+Xenses Lifestyle Spa & Salon

Aveda+Xenses 发廊及水疗中心

Hong Kong, China
中国香港

Area : 1,207.7m²
Completion Year : 2007
Design : Alex Choi
Photographer : Rage Wan

This salon-cum-spa is located in the basement of a shopping mall in Causeway Bay, giving it a natural edge over more exposed locations, with all thought of the hustle and bustle outside receding as visitors descend the purposefully tranquil entrance staircase. At the bottom, the visitor is greeted by a cool, predominately white-washed interior, muted by blocks of warm earthy tones. Contrast has been replaced by a colour and textural palette specifically chosen to create a calm and harmonious ambience, encouraging the clientele to settle back and relax in this unexpectedly peaceful setting. Because of its basement setting, the space was interrupted by a large number of columns that the designers sought to disguise by integrating them into a strictly rectilinear layout. Every element of the design is informed by this strong sense of geometry, from the rectangular blocks of colour on the walls to the rugs on the floor and much of the furniture, including the desks, tables and seating. As a result the columns have disappeared seamlessly into the overall composition, its strict angularity offset by subtle natural touches in the form of striking floral arrangements. And in the hair-dressing salon, a screen of reeds and curling branches — an exquisite work of art in its own right.

这间发廊兼水疗中心位于铜锣湾一间购物商场的地库层，有别于其他比较瞩目的地铺位置，当客人沿着楼梯进入中心时，犹如将街上的繁喧过滤，踏入别有洞天的静谧氛围之中。楼梯尽头为一个全白色的空间，泥土色调散落其中，颜色和材质互相协调，营造出一个舒泰和谐的环境，有助于客人放松身心，尽情享受中心所提供的各种服务。由于地库层布满大型的结构柱，我们采用掩眼法，将其融入长方形的平面布局之中。所有设计元素均贯彻着强烈的几何概念，如长方形的颜色墙，以至地板上的地毯，甚至办公桌、台面及座位等，令柱子仿如隐身在整个构图里。方正的线条被大自然花卉缀饰，发廊里更放有一面以芦苇及树枝勾勒而成的屏风，俨如一尊由大自然赏赐的艺术品。

2007 | Aveda + Xenses Lifestyle Spa & Salon

Aropa Mediterranean Restaurant

Aropa 地中海餐厅

K11 Art Mall, Hong Kong, China
中国香港 K11 购物中心

Area : 157.94m²
Completion Year : 2010
Design : Alex Choi
Photographer : Lam Ka Ho

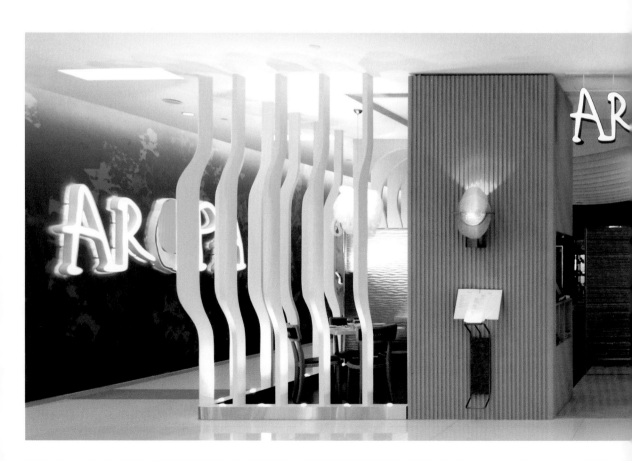

2010 | Aropa Mediterranean Restaurant

The idea to bring up the restaurant is based on the spirit of Anutans – "Aropa", which means giving, sharing, respect the nature. Located at the Solomon Islands, the island of Anuta stands out in one of almost a thousand islands, which is as pure as it has never changed since the creation of the world. The viewpoint from above Anuta gives a stunning panorama of the whole island. The wonderful panoramic view of the island of Anuta and the artistic handcrafts of the tribes become the inspiration of theme of the design. With the impression of characteristics of Anuta together with Aropa, the restaurant is shaped with plenty curves filling around from the side to the ceiling to create rhythmic rise and fall of the ocean. Like the island is in the breast of the ocean, the curvy open frames around the restaurant embrace the whole in an open way instead of isolation. The water jelly lamps and cloud chandelier are the highlights that stand out the concept of being natural. Materials inspired by nature are bringing in to this indoor area for the full representation in natural world and human nature. Besides enjoying the delicious meals, feeling the "Aropa" that will hold this busy city close to the nature.

建立餐厅的想法是基于阿努塔岛人的精神—"阿罗帕",这意味着给予、分享、尊重自然。位于所罗门群岛,阿努塔岛在几乎一千个岛中脱颖而出,它是纯粹的,因为自从世界诞生起,它就从未改变。上面关于阿努塔岛的观点给了整个岛一个令人震惊的全景。阿努塔岛美妙的全景和部落的艺术化手工艺品成为设计主题的灵感。在阿努塔与阿罗帕的特征影响下,餐厅塑造了大量曲线填充空间的侧面和天花板,创造出节奏性起伏的海洋景观。像这个岛位于海洋中间,绕着餐厅的弯曲开放方式框架以开放而非隔离的方式拥抱整个空间。水滴状灯和云状吊灯是重要的组成部分,突出自然的概念。受自然启发的材料引进室内,完整地表现了自然世界和人性。除了享受美味的饭菜,感受"阿罗帕",仿佛这个繁忙的城市都在亲近着大自然。

Suzuki Café chain
铃木珈琲馆连锁店

The One, Hong Kong, China
中国香港 The One 商场

Area : 111.48m²
Completion Year : 2011
Design : Alex Choi, Ian Leung
Photographer : Bobby Wu

Developing from the first design to Suzuki café chain in 2006, the design revamp had taken place in Hong Kong again. Although termed a café/restaurant chain, the Suzuki cafés depart from the standard preconception of a Japanese establishment to evoke a more European feel. Most of the company's outlets are located in shopping malls where spaces are often confined and internalised. Seeking opportunity in these limitations, the designers decided to create spaces that were deliberately intimate, almost homely in feel. In this context, a wood seemed an apt choice of material, and it was employed liberally throughout the various spaces, as walls, floors, screens and furnishings.

延续了设计师于 2006 年初次为铃木珈琲馆而定的设计概念,这次的设计改造同样发生在香港。铃木珈琲馆虽为一间连锁咖啡室/餐厅集团,然而设计上却摒弃了日式餐厅的传统形象,反而趋向欧洲化。集团大部分餐厅均设于购物商场内,空间狭小且封闭,面对这些限制,设计师决定塑造一种亲切、恰如家居的空间氛围,以原木覆盖空间内所有元素,包括墙壁、地板、屏风及家具。

| 2011 | Suzuki Café chain |

Caffè HABITŪ chain

Caffè HABITŪ 连锁咖啡厅

Hong Kong, China
中国香港

Completion Year : 2007 - 2008
Design : Alex Choi
Photographer : Rage Wan

Refreshing, sophisticated yet easy-going, caffè HABITŪ is a far cry from other chain coffee shops with their all too often bland identities and identical interiors. Here, the repeated use of a single design concept has been banished in favour of interiors that responds directly to their location and target clientele. If there is a common thread linking the various outlets, it lies in the use of abandoned materials left by previous occupants of the spaces. These might include wooden shelves reused as ceiling panels, or existing metal fittings and raw brick walls, while elsewhere PVC pipes might be transformed into ceiling lights or walls adorned with wooden shoe horns, creating sculptural features that imbue each space with a sense of its own history. Both the Henessy Road and Hutchison House outlets cater to office executives, but their different floor plans have dictated quite different seating layouts. In the former, the numerous twists and turns of the boundary walls have been employed to create secret corners, perfect for a private coffee break away from the office. In the latter, armchairs and sofas have been arranged around a central bar, in cosy clusters suitable for informal business meetings. Clad in copper mosaic tiles and highlighted by a rectangular black ceiling rimmed with spotlights, the bar was designed to encourage mingling and interaction, with sufficient visual impact to draw in passers-by even from a distance.

跟其他毫无新意的连锁咖啡厅截然不同，caffè HABITŪ 感觉既清新、优雅，又随意率性，设计师弃用了单一的设计词汇，让空间随分店的不同位置和目标客人来更换各种面貌。唯一贯穿各间分店的元素，便是前租客所遗留下来的物品，例如用木层架来铺砌天花，基本的金属制品和砖墙被保留下来，PVC 喉管被改造成天花吊灯，墙壁则饰以木鞋模，它们形形色色的仿如雕塑品，为空间添加了厚重的历史感。轩尼斯道及和记大厦两间分店的主要顾客为办公室行政人员，然而由于平面布局不同，座位的安排亦大相径庭。前者利用墙壁勾勒出隐密的角落，让客人可忙里偷闲；后者则设有一张中央吧台，四周围合的扶手椅和沙发方便客人一边喝咖啡，一边悠闲地聊天。吧台以闪亮的铜制马赛克包裹，长方形的天花四边安装了射灯，此安排不但鼓励客人彼此之间交流，更营造了别具一格的视觉效果，轻易地吸引店外途人的目光。

2007-2008 | Caffè HABITÜ chain

2007-2008 | Caffè HABITŪ chain

2007-2008 | Caffè HABITÜ chain

Caffè HABITŪ — the table

Caffè HABITŪ 咖啡店 —— 餐桌

Miramar Shopping Centre, Hong Kong, China

中国香港美丽华商场

Area : 167.22m²
Completion Year : 2011
Design : Alex Choi, Ian Leung
Photographer : Kit Siu

2011 | Caffè HABITŪ — the table

After the chain project of Caffè HABITŪ in 2007 to 2008, Caffè HABITŪ — the table is an Italian café that continuation of the former and it has taken the dining experience to an enhanced level. The Italian food culture is about sharing, people find a great pleasure in sitting at a table, in a home or restaurant and sharing a great meal together. The Table is intended to create a feel of Italian home kitchen, so setting a communal table at the centre of the space can evoke atmosphere and dining in here feels like joining in a meal at home. It is different from the authentic industrial feel in Caffè HABITŪ, a vintage touch and industrial chic elements have joined together to give the space a familiar and cozy feel.

在 2007 年至 2008 年的连锁项目咖啡 HABITŪ 之后，一家意大利的咖啡馆，咖啡 HABITŪ——餐桌，延续前者，将这个咖啡品牌提升到更高的水平。意大利的饮食文化是分享——人们觉得坐在家里或餐厅的桌子旁一起分享一顿美餐是一种很大的乐趣。这个餐桌的目的是创建一种意大利家庭厨房的感觉，所以要在空间中间创建一个公用餐桌，激起一种氛围——人们在这里用餐就像加入家庭聚餐般。它不同于 HABITŪ 咖啡的真正工业感觉，一种传统的感觉和别致的工业元素联系在一起，给空间一种熟悉而舒适的感觉。

RETAIL
零售

3 shop chain store
3 shop 移动电讯连锁店

Sole Town
Sole Town 鞋履配饰专卖店

EQ:IQ flagship store
EQ:IQ 服饰旗舰店

EQ:IQ chain store
EQ:IQ 服饰连锁店

Xenses cosmetic shop
Xenses 美妆品专卖店

GENIO LaMode men's fashion concept store
GENIO LaMode 男士服饰概念店

Marlboro Newstand
万宝路书报摊

Marlboro Counter
万宝路香烟柜台

K11 Design Store
K11 设计品味专卖店

3 Shop chain store

3 Shop 移动电讯连锁店

Hong Kong, China
中国香港

Completion Year : 2006 - 2010
Design : Alex Choi, Yan Yik Lun
Photographer : Rage Wan

2006-2010　3 Shop chain store

Brand identity is of crucial importance in Hong Kong's highly competitive mobile telephone environment, and 3 was keen to improve its market share with a new look for its shops that would command the customers' attention. The designers came up with a bold colour scheme of black and yellow that could be applied to all points of display, whether the various shop interiors or the product packaging, creating a distinctively co-ordinated look that immediately stands out. Contrasting with the glossy white floors and ceilings, the merchandise can be displayed against the colour that suits it best. Construction time also played an important role in the design and the desingers developed a system of wall panels that could be easily prefabricated off-site, allowing them to successfully revamp 50 stores in just two years.

在香港竞争激烈的移动电话市场中，品牌形象绝对是成败的关键。为了增加市场的占有率，3 shop 决定为店铺进行形象大改造，以吸引更多移动电话用户。设计师建议以黑和黄色作为主题颜色，并应用到店铺室内和产品包装上，令品牌的形象更一致、更鲜明。至于地板和天花则采用白色光漆，令陈列其中的产品更加突出。而建造时间亦是其中一个重要的考虑因素，设计师特别设计了一个独特的墙壁镶板系统，能够事先于工厂里预制，再运送到工地快速装嵌，令他们成功于两年内完成 50 间店铺的改造工程。

Sole Town

Sole Town 鞋履配饰专卖店

Hong Kong, China
中国香港

Area : 1,254.19m²
Completion Year : 2007
Design : Alex Choi
Photographer : Rage Wan

Sole Alliance is a multi-brand - covering shoes and accessories for today's women - run by Hong Kong - based GRI Group. Sole Town, the brand's new store in Hong Kong, is the work of Alexchoi design & Partners. The most well-known story based on 'a shoe' is surely the fairy tale featuring Cinderella and the Prince, which became the inspiration for the designers' retail concept; an exciting, energetic, youthful environment with a touch of elegance. What a dream like paradise for young girls. Silhouettes of birds and trees representing an enchanted forest decorate the windows of two retail areas - Sole Glam and Sole Hollywood - where customers find shoes by Jessica Simpson and Jennifer Lopez. A third zone illuminated in purple and blue, the sculptural Sole Cool area, is a metaphor for Cinderella's tenacious character. It's here that shoppers symbolically try on the famous glass slipper. Shiny surfaces and sparkling crystal chandeliers in the Sole Classy area evoke the royal palace. Marking Sole Chic are several display islands in an open space. A bold pattern of lines on the wall contrasts with Nine West pendant lights. Each zone, along with the scene it represents, accommodates a different brand and type of shoes. Continuous, streamlined linear shelving pulls the elements of the plot together, softens the angularity of the overall space and works to facilitate the circulation plan, leading customers through the shop and, ultimately, to the monolithic black steel-clad cash desk. Another binding element is zigzag-patterned flooring, a wood surface appearing in most areas of the shop.

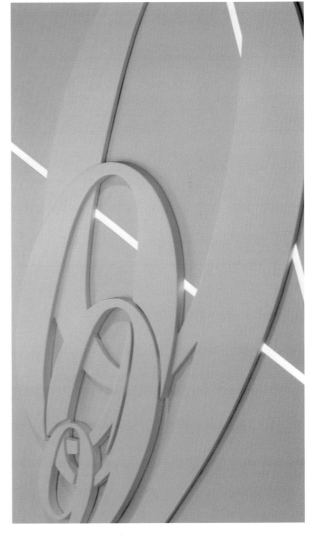

鞋履是一间时尚鞋履与包款配件的复合型品牌概念店，由香港GRI集团运营。Sole Town是秀履在香港开设的新店，由Alexchoi design & Partners设计。关于鞋子最广为人知的是灰姑娘的故事，新店的设计理念就从其中汲取了灵感，打造出一个令人兴奋，有活力，有朝气的优雅空间。这就是一个年轻女孩的梦中天堂。森林风格的鸟，树轮廓点了两个品牌——魅履品牌和好莱坞品牌的玻璃展示橱柜，前者是杰西卡·森普生设计，后者是詹尼佛·洛佩兹设计。另外一个品牌区展示了酷履系列，紫蓝色的灯光营造出别样感官，是灰姑娘倔强个性的隐喻。在这里，顾客可以象征性地试穿灰姑娘的水晶鞋。经典系列的品牌区装置了一盏闪亮的水晶吊灯，营造出一种尊贵感。开阔空间的展示台为潮流展示量身定做。墙面的大胆线条图案与Nine West吊灯形成鲜明的对比。每一个品牌区都采用了不同场景设置，展现了多种鞋类品牌和设计。连续的线形展示架将所有元素连接在一起，柔化了整个空间的棱角，同时引导了人流走向，顾客沿着货架穿过商店最后到达黑钢覆盖的收银台。另外一个模糊空间界限的元素是铺设了大部分店内地面的曲折图案的地板。

2007 | Sole Town

EQ:IQ flagship store

EQ:IQ 服饰旗舰店

APM Beijing, China
中国北京新东安广场

Area : 389m²
Completion Year : 2008
Design : Alex Choi
Photographer : Ken Choi

EQ:IQ is an apparel, accessory and home-decoration brand belonging to the Hong Kong - headquartered in GRI Group. Launched in 1999, the brand sets its target on modern, confident and well-traveled women are age 25 to 55. Designer transferred the characteristics of the EQ:IQ brand to the interior of a new shop in Beijing. The designers aimed for a concept that - much like EQ:IQ fashion collections - is minimalist in style but rich in detail and texture. They divided the U-shaped store into three retail zones. Accessible from the main entrance is an area where the most important pieces of the current collection can be found. A custom-made chandelier, which immediately grabs the customer's attention, consists of a cluster of rods made from solid ash, acrylic resin, brass, and metal painted matte white. Below the chandelier is a rug featuring the EQ:IQ branding icon: a tree. Farther along and to one side, a narrower area contains fashions from the young-women's collection, which are showcased on transparent shelves supported by vertical brass tubing across from and contrasting with - a horizontal pine structure that screens off the fitting area. Shoes and bags are displayed in the third zone, which has another point of access: the entrance to the store from the shopping mall. A nearby escalator inspired the white staircase in the shop window, a tiered element that serves as a product display. The exterior of the shop is wrapped in a tree pattern that unifies the whole. High-intensity discharge downlights give the store a sparkling appearance after dark.

EQ:IQ 是属于香港的服装、配件、家居装饰品牌——总部设在 GRI 集团。成立于 1999 年，品牌的目标客群设定为现代、自信和见多识广的 25 至 55 岁女性。设计师把 EQ:IQ 的品牌特点转化在一间在北京的新店。旨在为一个概念，就像 EQ:IQ 的系列时装一样，简约但充满丰富细节和质感。我们把 U 形店面分为三个零售区。从正门踏入店中，便可发现当季最新系列的服饰。抬眼顿时叫人驻足，一盏由一束实木水曲柳、树脂、铜、哑白焗漆金属棒组成的定制装饰吊灯在店内璀璨。吊灯之下是印有 EQ:IQ 品牌标"树"的特色地毯。沿路来到店的另一侧，一个小区域内集结了时尚年轻女性的系列服饰全部展示在垂直黄铜管透明层架上，与对面把试衣间分隔开的松木水平结构形成强烈对比。鞋子和包包在第三区，此区有一个从商场到店内的入口。附近的一个自动扶梯成为了商店橱窗白色楼梯的设计灵感，一层层阶梯元素换成商品陈列架。整个店面被包裹在树的图案里，营造整体感觉。入夜后在射灯的映照下，整个店面比白天更加鲜明。

2008 | EQ:IQ flagship store

EQ:IQ chain store
EQ:IQ 服饰连锁店

China
中国

Completion Year : 2008-2010
Design : Alex Choi, Marco Choi
Photographer : AlEXCHOI design & Partners

The corporate logo of EQ:iQ is a tree, a motif that the designers decided to reference in all the company's stores — sometimes boldly as large murals on otherwise plain walls, and sometimes more subtly in the form of loose rugs on the floor or as suspended elements in the space. In the company's clothes stores, a mixture of austere black and softer white-painted walls is punctuated by pale wood and sandblasted glass panels, to create a neutral backdrop against which individual garments can stand out. In the same spirit, mannequins and merchandise are kept to a minimum so that the items on display can be appreciated from every angle. In keeping with the neutral background, shelves, tables and fittings have been kept equally simple, though a garment rack, resembling a luggage trolley found in smart hotels, offers interest with a light touch.

EQ:IQ的企业标志是一棵树，设计师决定依循这个图案，贯彻到公司旗下的每一间店铺内，如大型壁画、小型地毡，或是空间里悬吊着的元素。所有服装连锁店均以黑、白墙壁为主调，间或加进点点原木和磨砂玻璃屏板，营造出一个柔和的中性背景，令陈列其中的服装更形突出。就连模特儿公仔和货物的数目也不多，让客人可以用不同角度欣赏陈列品。货架、桌面和其他装置都以中性背景为前提，设计简朴无华，而其中一个形同酒店行李手推车的衣物挂架，低调地添上一抹生动的笔触。

2008-2010　　EQ:IQ chain store

2008-2010 | EQ:IQ chain store

2008-2010　EQ:IQ chain store

Xenses cosmetic shop

Xenses 美妆品专卖店

Hong Kong, China
中国香港

Area : 120m²
Completion Year : 2007
Design : Alex Choi
Photographer : Rage Wan

FRAGMENTATION = FOCUS

Brand identity is usually of utmost importance in retail design, but for Xenses any sense of definitive style has to be avoided as, rather than displaying its own products, the space is used to sell a broad range of skincare and cosmetic brands by others. To demonstrate this multi-brand nature, six different colours of tinted glass, cut into rectangles of varying sizes, were chosen to clad all the walls and columns. Seen from afar, they form an intriguing patchwork pattern that reflects and softens the recessed lighting as well as helping to hide the columns that interrupt the space. These were further disguised by enlarging them so that niches could be inserted for display purposes, this interplay of solid and void being used throughout the space to exaggerate the visual complexity and enliven an otherwise austere retail environment.

在零售设计的层面上，品牌形象就是王道；然而 Xenses 的设计却避免采用单一风格，因店铺主要售卖其他品牌的一系列美容及化妆产品，而非自家出品。为了展示这种多元品牌本质，设计师采用了六种不同颜色的玻璃，切割成不同大小的长方形，包裹着所有墙壁和柱子。从远处望去，它们形成有趣的拼贴图案，不但柔化了嵌灯的光线，同时隐藏了穿插于空间中的柱子。这些柱子更被掏空作陈列用途，虚与实的对比贯穿整个空间，加强了视觉上的错综层次，为原本简朴的商店环境增添趣味。

GENIO LaMode men's fashion concept store

GENIO LaMode 男士服饰概念店

Hangzhou, China
中国杭州

Area : 232m²
Completion Year : 2011
Design : Alex Choi, Yan Yik Lun
Photographer : Alex Choi, Yan Yik Lun

GENIO LaMode is a new menswear fashion label which is mass-oriented in mainland China. The brand has aspirations to start a fashion campaign in the competitive market while the designer has the aim to evoke the spirit of smart casual in this fashion showroom. The fashion of smart causal is not a comprehensively explored field at that time. In stead of creating an ordinary showroom that displaying clothes, how to present the attitude of smart and causal became the main focus of the showroom. Black iron and pine wood had made a huge contract to the interior that brought in the exotic ambience and had interpreted the balance between work and playful. The concept becomes visible in encounters with contrasting materials, the profusion of displays and racks help to stand out the variety of collections while a system is hidden. The unique design system is made for the mass roll-out to China and even Asia. Although the design is one of its kind in the market thus far, but it is also compatible to the mass market. At the end, fashion is all about presenting your own personality and the designer translated the message into the interior.

意森是一个新的男装时尚品牌，面向中国大陆的广大消费群体。该品牌有志于在竞争激烈的市场中开创新的时尚潮流，同时设计师也希望打造出一个有着休闲装精髓的时装展销店。当时休闲装的时尚尚未兴起。该时尚展销店的焦点并非打造一个陈列衣服的普通店面，而是如何体现轻松随意的态度。黑钢和松木与室内形成了鲜明对比，带来了异国的氛围，并诠释怎么样在工作与娱乐之间取得平衡。这个概念使用了有对比的材料，变得视觉化，丰富的展示和挂物架使得服装脱颖而出，系统则变得隐蔽。独特的设计系统是为该品牌在中国甚至亚洲的大规模上市而打造。虽然迄今为止，该设计只是它在市场上的类型之一，但也符合大众市场。最后说来，时尚都是关于展示自己的个性，设计师则将这融入了室内。

Marlboro Newstand
万宝路书报摊

Hong Kong, China
中国香港

Completion Year : 2006 - 2010
Design : Alex Choi, Yan Yik Lun
Photographer : ALEXCHOI design & Partners

2006-2010 | Marlboro Newstand

Hong Kong's ban on tobacco advertising in public places has prompted Marlboro to come up with innovative ways of branding. The result is an unconventional newstand incorporating a small secure display of Marlboro cigarettes, but otherwise downplaying the branding to just the subtle use of narrow strip lighting in the company's corporate colour, red, and the colour can also be changed seasonally to accommodate marketing purpose. Manufactured in rugged, anodised steel panels, the kiosk folds out from a relatively small sealed box, when closed, to create a variety of vertical display surfaces, as well as a large single-tiered platform on which a generous array of magazines and newspapers can be displayed to encourage free browsing. LED lighting keeps electrical power requirements to a minimum, while clear plastic screens installed in the canopy section can be rolled down in the event of rain.

香港政府有关禁止烟草广告的规定，促使烟草商 Marlboro 构思更新颖的方式以推广品牌形象。最佳例子莫过于本个案的报纸摊档，报摊除了设有香烟的陈列饰柜外，更以低调的手法重新演绎品牌的企业颜色——红色。整个报摊由电镀铝板拼合而成，边缘缀以一条条红色幼灯管，在黑夜的街道上煞是抢眼，灯管的颜色还可以随时改变以配合市场推广。当铝板合上时，整个报摊无异于一个黑色盒子；打开铝板后，一个个垂直陈列架顿时出现，其中的平台更可展示大量杂志和报纸，让顾客一目了然。报纸摊档以 ERD 作照明，可将耗电量减至最低，而顶篷位置亦细心地加设透明塑料屏风，以防下雨天时弄湿报刊。

Before

2006-2010 | Marlboro Newstand

Before

Open

Close

2006-2010　Marlboro Newstand

After

Marlboro Newstand | NO.76 - 77

Marlboro Counter
万宝路香烟柜台

Hong Kong, China
中国香港

Completion Year : 2006 - 2010
Design : Alex Choi, Yan Yik Lun
Photographer : ALEXCHOI design & Partners

The Marlboro counter is another take on the news stand, albeit smaller in scale and with a simpler programme. In the news stands the cigarette display is placed well out of reach at the back of the kiosk, but as a freestanding unit it can be brought forward and inserted into any number of suitable locations. Recognising that the counters would most often be placed in the small family-run convenience stores found all over Hong Kong, we deliberately chose sleek black surfaces and sharp lines of red LED lighting as the primary finish for the units, to create a striking juxtaposition with the surrounding stacks of snacks and magazines - a small corner of luxury which is sure to capture the customers' attention and fulfil its intention as a branding exercise.

万宝路香烟柜台是报纸摊档的另一个变奏，规模和内容都相对地简单。报纸摊档基于安全原因，往往将香烟饰柜设于收银员的背后位置，然而陈列柜台却可置于任何显眼的位置。这些香烟柜台多设于香港各区的家庭式商店里，设计师因此选用了黑色背景和红色 LED 灯管作为主要的设计元素，令香烟柜台与商店里摆放的小食和杂志形成强烈对比，形成一个亮眼的角落，能够于瞬间吸引路人目光，成功建立起鲜明的品牌形象。

2006-2010 | Marlboro Counter

K11 Design Store

K11 设计品味专卖店

K11 Art Mall, Hong Kong, China

中国香港 K11 购物中心

Area : 2,560m²
Completion Year : 2011
Design : Alex Choi, Stephanie Leung
Photographer : Kit Siu

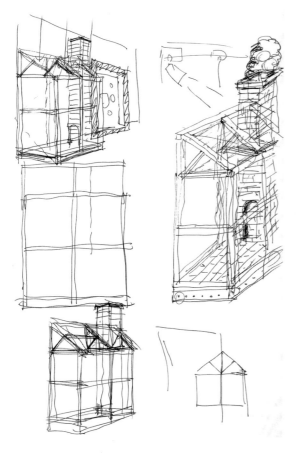

K11 retail concept store is a popular lifestyle department store at Tsim Sha Tsui in K11 shopping mall selling furniture, lighting, designers' items and home decor accessories. Most of the products are playful, stylish designer items selectively chosen from foreign country. Designer uses the "contemporary gallery" concept to display every unique and inspirational items, like setting up slot board flexible display system. Flexibility is important to running this special department store. In addition, materials selected is to create a home-like feeling by using tile clad finish in brick pattern, tile clad overhang wall panel, teak wood and olive green, just like the retro bathroom or kitchen. People can easily find their lifestyle by choosing their unique items in this mix & match home.

K11 Design store 是尖沙咀 K11 商场里的一家受欢迎的关于时尚生活的百货店,出售家具、照明、设计器具和家居装饰用品。大部分的产品都是精选的国外好玩而时尚的设计师物品。设计师使用了"当代画廊"的概念来展示每一个独特的和鼓舞人心的项目,比如说它有一个灵活的槽板展示系统。灵活性对于这个特殊的百货商店的运作很重要。另外,它在材料选择上通过使用带图案的抛光瓷砖、瓷砖复合墙壁面板、柚木木材和橄榄绿打造一种宾至如归的感觉,宛如复古的浴室或厨房。人们可以在这个混搭风格的店中筛选独特的商品,轻而易举地找到自己的生活方式。

2011 K11 Design Store

OFFICE
办公室

Jeanswest International (H.K.) Ltd. office
真维斯国际(香港)有限公司办公室

ALEXCHOI design & Partners office
蔡明治设计有限公司办公室

Jeanswest International (H.K.) Ltd. office
真维斯国际(香港)有限公司办公室

Hong Kong, China
中国香港

Area : 940m²
Completion Year : 2008
Design : Alex Choi
Photographer : ALEXCHOI design & Partners

Transparency is the first priority of the design. It is a precious opportunity to have a site like this which can enjoy the "unique" spots of Hong Kong, like the old Kai Tak Airport and the Lion Rock. Clear glass full height partition with elaborated and minimal stainless steel fixing composed the "Air" like senior corners. It blends with the open planned general staff area. Blue and green colours are used to divide the subsidiaries of Jeanswest. It starts from reception in a symmetrical arrangement. It created a clear definition of space but consistent. The colours lead to the heart of the staff area. White color system furniture then is the only option in order not to create competition with the feature colors. Slick and sophisticated details help to enrich such a linear and minimal space. Product development section is the most important part of Jeanswest, so there are two brainstorming rooms with totally different "Look & Feel". Custom designed and stainless steel made hanging racks make the rooms creative and practical. A transparent showcase is designed to display their new invention of fabric material. And the allocation along the main aisle intends to present the "Forward" & "Pioneering" of Jeanswest philosophy.

我们的设计中,"透明"是首要原则。能在香港欣赏到如旧启德机场、狮子山这些独特的景致,实在是一个非常宝贵的机会。透明玻璃和不锈钢共同组合成如空气般清新的高级管理人员区域,其他开放式空间为普通员工办公区。真维斯公司附属机构的分区则采用蓝色和绿色。首先是接待室,清淡的用色衬托出一片清新的空间,而颜色的均匀分布又使得空间的风格高度统一。蓝绿两色继续延伸到员工区域的心脏地带。为了不发生颜色上的冲突,办公设施选择了白色,完美融合于整体主题色系当中。另外平滑精致的细节设计能起到拓宽空间的作用。产品开发部是真维斯最重要的部门。为此设计师专门开辟了两间外观和感觉上完全不同的讨论室。采用了传统设计手法的不锈钢架使得房间既充满创意又脚踏实地。透明的陈列橱可用来展示公司开发的最新布料。过道两旁的设计体现了真维斯"前进、开拓"的公司理念。

| 2008 | Jeanswest International (H.K.) Ltd. office |

ALEXCHOI design & Partners office

蔡明治设计有限公司办公室

Hong Kong, China
中国香港

Area : 325.16m²
Completion Year : 2008
Design : Alex Choi
Photographer : Rage Wan

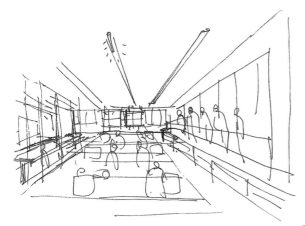

Located in a 70s modernistic industrial building which the structure is a complex of thick columns, beams and ribs. In order to break the rigidity, a swish-like curve is introduced at the reception which brings the incomers to the first core structure — meeting room. The cylindrical structure acts as a central point, with the office working area at the north of it, pantry and washrooms at the west, conference room at the northwest and exit at the south. It provides a sense of orientation. The main curve aisle is also a gallery-like space displaying the owner's collections — designer chairs. Other than the chairs, incomers will be amazed by another collection — a gigantic bookshelf in curves. The intent of this arrangement derives from the owner's belief on running this design business, which is "inspirations are around you all the time". The curvy bookshelf partitions the recreation/conference room and leads to the working area. The linear workstations echo with the ceiling ribs structure which is enhanced by the linear up - lit tubes. The linear feature ends at the black framed windows which captures the exterior. The use of black and white colour reinforces the flow of space, just like putting charcoal lines on a white paper.

办公室位于七十年代时建起的现代工厂大厦内,厚实的柱子、横梁及拱壁使空间形成了复杂且刚直的结构。设计师以柔制刚,用妩媚的曲线把来访者从接待处引领到第一个中心点——会客室。为提升空间的方向感,布局以柱体的会客室为中心点,工作间被放置在办公室的北面,茶水间和洗手间处于以西的地方,而会议室和出口就分别位于西北及南面。流线型的主要通道亦是设计师展示其珍品——设计师椅子的空间。除此以外,访客必定会为弯弯的巨型书架而惊叹,这个设计灵感是源于设计师对设计的理念:灵感随处可见。书架把休憩空间和会议室分隔开并延展到工作间;直线排列的工作间与横梁及天花的仰光灯互相呼应,直至黑框窗户为终点。黑与白的衬托加强了空间的流动性,随性的布局就像白纸上的素描线。

2008 ALEXCHOI design & Partners office

RESIDENTIAL
住宅

Apartment in Harbourside, Hong Kong
香港君临天下住宅项目

Apartment in Harbourside, Hong Kong

香港君临天下住宅项目

Hong Kong, China
中国香港

Area : 112m²
Completion Year : 2007
Design : Alex Choi
Photographer : Alex Choi

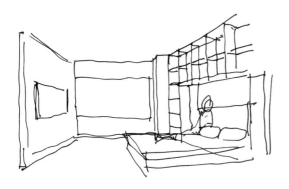

This apartment is designed for a young family of three, who wants a simple and cosy home to go back to after hectic hours. Just as its name implies, it is located in Hong Kong that next to the very famous Victoria Harbour. With the great harbour view setting outside at the middle of the apartment, there will be no other substitutes to replace this vivid art painting as the focal point of the interior. Simply using white as the main tone of the whole area while sprinkling wood element to define the space. There is no extra decoration, the apartment stays simple and cosy, but also sophisticated and contemporary.

这个住宅是为一个年轻的3口之家而设计，他们需要一个可以让他们结束繁琐的一天后，回去的简单而温馨的家。正如住宅的名字暗示，它坐落于香港非常著名的维多利亚港旁边。在住宅的中央，壮丽的海港景色就这么置于窗前，以这帧生动的画作作为重点的内部是没有其他替代品可取代的。以简单的白色作为整个区域的主基调再加入木的元素来把空间划分开。没有多余的装饰，这个家简单、舒适，也是精致和充满现代主义的。

2007 | Apartment in Harbourside, Hong Kong

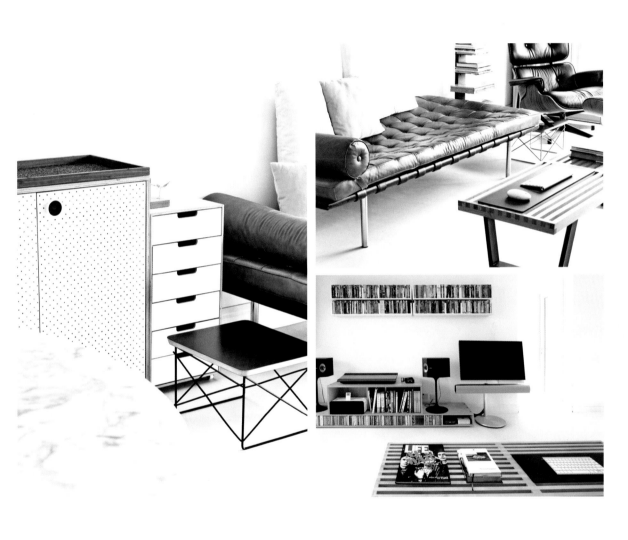

| 2007 | Apartment in Harbourside, Hong Kong |

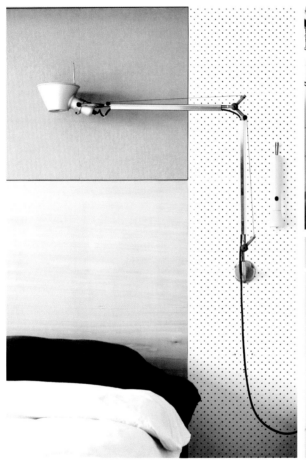

Apartment in Harbourside, Hong Kong

Acknowledgements

We would like to thank all the designers and companies who made significant contributions to the compilation of this book. Without them, this project would not have been possible. We would also like to thank many others whose names did not appear on the credits, but made specific input and support for the project from beginning to end.

Future Editions

If you would like to contribute to the next edition of Artpower, please email us your details to: artpower@artpower.com.cn

ALEXCHOI
design & Partners

collections
蔡明治设计精选作品集

Volume Three
Definitive
第三册 坚守原创

深圳市艺力文化发展有限公司 编

华南理工大学出版社
SOUTH CHINA UNIVERSITY OF TECHNOLOGY PRESS
·广州·

Keeping your own style & uniqueness in the market is the best way to stay alive.

SINCE 1997

DEFINITIVE

———————————————————————————— SINCE 1997

图书在版编目（CIP）数据

蔡明治设计精选作品集 = ALEXCHOI design & partners collections.3，坚守原创 / 深圳市艺力文化发展有限公司编 . — 广州 ：华南理工大学出版社，2014.4
ISBN 978-7-5623-4118-5

Ⅰ．①蔡… Ⅱ．①深… Ⅲ．①室内装饰设计－作品集－中国－现代 Ⅳ．① TU238

中国版本图书馆 CIP 数据核字（2014）第 059402 号

蔡明治设计精选作品集 ALEXCHOI design & Partners collections
第三册 坚守原创 Volume Three Definitive
深圳市艺力文化发展有限公司 编

出 版 人：	韩中伟
出版发行：	华南理工大学出版社
	（广州五山华南理工大学 17 号楼，邮编 510640）
	http://www.scutpress.com.cn E-mail: scutc13@scut.edu.cn
	营销部电话：020-87113487 87111048（传真）
策划编辑：	赖淑华
责任编辑：	陈　昊　赖淑华
印 刷 者：	深圳市汇亿丰印刷包装有限公司
开　　本：	787mm×1092mm　1/16　**印张：** 22.75
成品尺寸：	185mm × 260mm
版　　次：	2014 年 4 月第 1 版　2014 年 4 月第 1 次印刷
定　　价：	398.00 元（共 3 册）

版权所有　盗版必究　　印装差错　负责调换

CONTENTS

ENTERTAINMENT
娱乐

Broadway Cinema chain　　　　　　　　2
百老汇影城连锁影院

Palace Cinema chain　　　　　　　　　10
百丽宫电影城连锁影院

Jinyi International Cinema　　　　　26
金逸国际影城—上海龙之梦

HOSPITALITY
餐饮休闲

Greyhound Café　　　　　　　　　　　34
Greyhound 餐厅

Greyhound Café　　　　　　　　　　　40
Greyhound Café 餐厅

Crystal Jade La Mian Xiao Long Bao　46
翡翠拉面小笼包餐厅

Crystal Jade La Mian Xiao Long Bao　50
翡翠拉面小笼包餐厅

RETAIL
零售

3 Mobile flagship store　　　　　　　56
3 移动电话旗舰店

3 Mobile chain store　　　　　　　　 60
3 移动电话连锁店

Hang Seng Bank retail branch revamp　64
恒生银行分行形象改造

Garden of Eden　　　　　　　　　　　76
伊甸园

Leo Optic eyewear　　　　　　　　　 80
护视光学眼镜店

TAG Heuer concept store　　　　　　 84
TAG Heuer 概念店

Moment fashion watch concept store　86
Moment 时尚手表概念店

ARTĒ Madrid concept store　　　　　 88
艾尔蒂概念店

Watson's Wine Cellar chain store　　92
屈臣氏酒窖连锁店

Fook Ming Tong Tea shop　　　　　　 94
福茗堂茶庄

Arome bakery room chain　　　　　　 96
东海堂连锁饼店

Dymocks book store　　　　　　　　　102
Dymocks 书店

OFFICE
办公室

Royal Spirit office　　　　　　　　 108
Royal Spirit 办公室

The ROOM studio　　　　　　　　　　 116
The ROOM 工作室

ENTERTAINMENT
娱乐

Broadway Cinema chain
百老汇电影城连锁影院

Palace Cinema chain
百丽宫电影城连锁影院

Jinyi International Cinema
金逸国际影城—上海龙之梦

Broadway Cinema chain
百老汇影城连锁影院

Hong Kong, China & China Mainland
中国香港及中国内地

Design : Alex Choi, Yan Yik Lun, Lewis Ho, KK Chen
Photographer : Guozeng Jiang, Bobby Wu, Kit Siu, Apollo So

2010-2012 | Broadway Cinema, MIXc Hangzhou, China

Broadway cinemas has be taking the lead of the industry for almost a decade. Their chain cinemas strategy spread not only in Hong Kong, China but also in China Mainland. For the broadway cinemas chain design projects, the designers have made a new definition to "chain design" — They create prototype, and roll out varieties.

The broadway cinemas were in Hangzhou, spanning four levels inside a high-end shopping mall. The twists and turns between its rooms served as inspiration for designers, and created differing aesthetics level by level. Outside of the theatres, space is bright and white; inside, dark walls and muted tones emphasize the ambiance of film watching.

The broadway cinema in Hong Kong Plaza Hollywood is designed to be chime with the overall ambience of the shopping mall, but in an extraordinary way. The designers decided to have the wooden element as the theme of the cinema and it enhances the warmth throughout the space. The complexity of the colour usage of orange and red in the theatres defined the vibrancy of the whole cinema and evoked the excitement movies.

The 402m^2 in Hong Kong Cyberport broadway cinema is evenly divided into 2 sections — "Excitement" and "Chilling". The outlines of the custom made perforated metal wall panel creates a cloud wave, it leads all the way through the cinema. Beside the snack & wine bar, a chilling area was put together between the theater entrances and the bar. Yuppies can hang out a bit before they have movies or just laying back with friends.

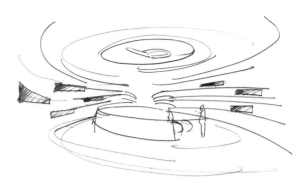

百老汇影城已经引领这个行业十多年了。他们的连锁策略不仅风行中国香港，在中国内地也在迅速扩张。如今百老汇影院的连锁理念有了新的定义。设计师创造一个原型，但在原型之上可以各自发挥。

杭州的百老汇影院位于一座高端购物商场之中，上下共4层。不同空间的迥异细节激发了设计师的灵感，每一层都蕴含着不同的美感。剧院外的空间明亮通透，剧院内的暗色墙壁和哑色色调却强调了观影氛围。

香港荷里活广场的百老汇影院的设计理念是要符合购物商场的整体氛围，同时要独具特色。设计师决定使用木元素作为影院的主题，提升整个空间的温馨感。橘色和红色的交替使用体现了影院的活力，使得影片更有冲击力。香港数码港的百老汇影院共有402m^2，这个空间被平分成两个部分——一个空间充满兴奋感，一个空间让人安定。定制的打孔金属壁板创造出一片波浪云，环绕了整个影院。餐饮吧的旁边，就在影院入口和吧台中间是一个休闲区，雅皮士们可以在影片放映之前在此休息片刻。

| 2010-2012 | Broadway Cinema, Cyberport, Hong Kong |

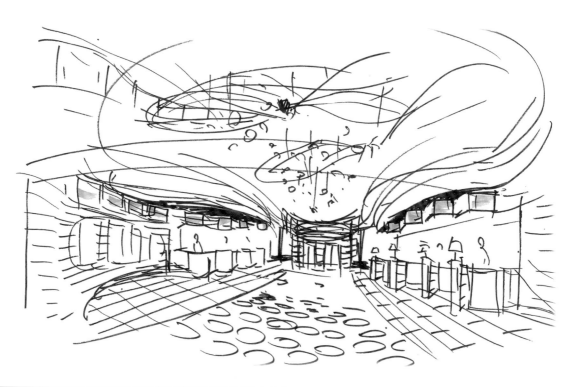

2010-2012 | Broadway Cinema, Plaza Hollywood, Hong Kong

Palace Cinema chain
百丽宫电影城连锁影院

Hong Kong, China & China Mainland
中国香港及中国内地

Design : Alex Choi, Marco Choi, Lewis Ho, KK Chen
Photographer : Bobby Wu, Kit Siu, Guozeng Jiang, Apollo So

2009-2013　Palace Cinema, ifc, Hong Kong

As the pioneer of the cinema industry in Hong Kong, broadway cinemas has their spots spread in almost every shopping malls in city. The last five years, with the film industry is heating up in China, broadway cinema wants to stay ahead in the industry, so setting foot in all first-tier cities of China is a significant step that needed to be taken.
The so called "prime locations" in China always pick up their pace to get close to the global trend and takes the lead to show how metropolitans enjoy lives. However, its uniqueness has not been washed out by trend, those cites' characters remain and all the cinemas adopted chic boutique style that comprised with the complexity of using bronze and gold has beautifully emphasized the layers. Extract from the uniqueness of the cities with the theme of classic with twist, different signatures were applied to different cinemas.
We brought a new razzle-dazzle from the dream factory by decorating the Palace ifc with arches, and furnishing it with different tone and texture of metal. This lifestyle cinema is extraordinarily modern in its function. A lifestyle book store and café feed those sophisticated movie-goers not only with movies but also words and delicate cuisine. The oversized conspicuous light frames in the Shenzhen broadway with classical type face slow down the passers-by's fast pace novel enough to catch their eyes. To recall audiences' curiosity and excitement of the movie, we put on curtains for the cinema in Chengdu. The natural of Jinan's (as known as "City of Springs") beauty is the inspiration for us to design the cinema in Jinan as a water palace. The walkway along the theaters is shaped into an undulating outline to imitate the stream that flows through the cinema while the metal mosaic gives the space a shimmering, wavering appearance. The dreamy cloud-like feature ceiling has expanded the infinite imagination through out Palace cinema in Shenyang. Exaggerate polygon design with the use of high contrast black and white colour, the impression of this cinema in Shanghai iapm is breathtaking and this edgy design gives broadway cinema a new energy in terms of a chain design.
Although these cinemas have their own feature, they are not separating from each other and remain as a whole chain of cinema when keeping the core style. Keeping your own style and uniqueness in the market is the best way to stay alive and this is the signature move for broadway to dance on the stage of a great country like China.

作为香港的行业先锋，百老汇影院的身影遍布了几乎所有购物商场。在过去的五年时间，电影行业在中国不断升温，百老汇影院希望能始终屹立于行业顶端，所以进军中国一线城市是至关重要的一步。
所谓的黄金地段总是迎合全球潮流趋势，敢于人先地展示都市生活的活色生香。然而这并没有消除它的独特性。城市的鲜明特色得到了保留，所有的影院都采用了精品潮流设计，融合铜与金的使用，体现出空间的层次感。从城市特色提取的经典元素与别具一格的设计结合创造出各具风情的影院。不同颜色和质地的金属装点了从梦工厂借鉴的新花样——拱顶。风格化的书店和咖啡店为观影者提供了文学与美食。深圳百老汇影院就装置了超大的灯架，路人经过都放慢脚步。为了唤起观影者的好奇心与兴奋感，我们在成都的百老汇影院装上了幕帘。被冠以泉城之称的济南极富自然美，这给了我们灵感。于是我们将济南百老汇影院设计成一座水的宫殿。沿着剧院的走道是波浪状的，模仿从影院流过的水流，金属马赛克又增添了一种动态感。沈阳的百老汇影院的天花板仿若云朵，激发无限想象。上海百老汇影院采用多边形设计与对比强烈的黑白色，令人叹为观止，这种潮流的设计也为百老汇连锁注入新的活力。
虽然这些影院都有各自的特色，它们并没有脱离彼此，作为连锁影院，它们仍然保持了一种整体性。与众不同是立足于市场的法宝，百老汇在中国的拓展正是这一规则的体现，也是企业拓展的一大步。

Palace Cinema, ifc, Hong Kong

2009-2013 | Palace Cinema, ifc, Hong Kong

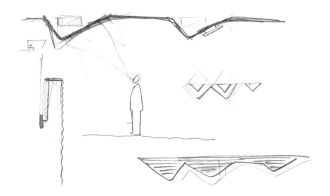

2009-2013 | Palace Cinema, Parc 66 Jinan, China

| 2009-2013 | Broadway Cinemas, Coco park, Shenzhen, China |

Broadway Cinemas, Coco park, Shenzhen, China

2009-2013　Broadway Cinemas, MIXc Chengdu, China

2009-2013 | Palace Cinema, Shenyang Forum 66, China

| 2009-2013 | Palace Cinema, iapm, Shanghai, China |

Palace Cinema, iapm, Shanghai, China

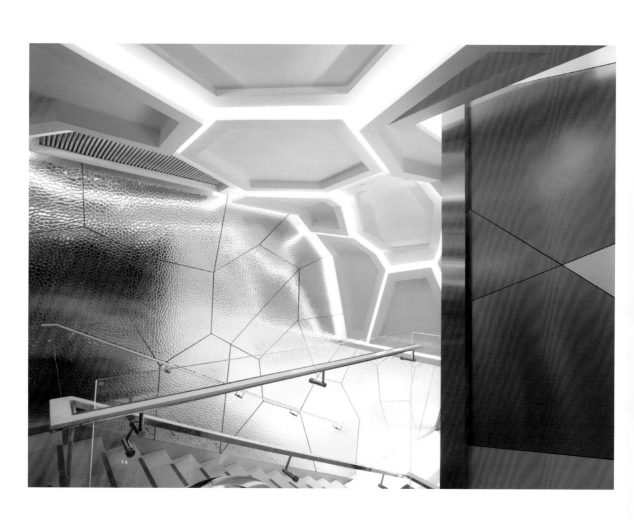

| 2009-2013 | Palace Cinema, iapm, Shanghai, China |

Jinyi International Cinema
金逸国际影城

Longemont, Shanghai, China
中国上海龙之梦

Area : 5,799.94m²
Completion Year : 2010
Design : Alex Choi, Marco choi
Graphic Designer : Kit Siu
Photographer : Guozeng Jiang

2010 | Jinyi International Cinema, Shanghai Longemont

If cinemas are places for dreamers, then its better interpretation should be the forest of dream. Located in the chic and classy shopping mall of Shanghai, which the name of the mall is also related to fancy dreams, the Jinyi International Cinema renders a feeling of being in a secret forest. Throughout the subtle entrance, it is an altogether different world from the outside. The 8.3 m ceiling height with the apposition of tree trunks, as well as boasting a concept in an amplify way by shredding the deformed leaves all over the ceiling and attaching to the trunks and walls. Together with the use of cool blue and illusory red create a dreamland where the film goers can chill out in the bird-nest-form movie book cafe or dream in the cocoon-like VIP room. The design of the cinema creates comfortable spaces and entices film goers from reality.

如果说电影院是梦想家的地方，那么它更好的解释应该是梦想的森林。位于上海别致而优雅的购物商场——商场的名字也与华丽的梦想有关，金逸国际影院呈现出秘密森林般的感觉。从精细的入口开始，它就是一个与外面完全不同的世界。8.3 m高的天花板与同样高的树干，以一种放大的方式夸大概念，即粉碎天花板上变形的树叶附到树杆和墙上。用很酷的蓝色和虚幻的红色创建一个梦境，影迷可以在鸟巢式的电影书咖啡馆冷静下来或在茧化蝶般的VIP房间幻想。电影院的设计创建了舒适的空间，并吸引影迷从现实进入电影。

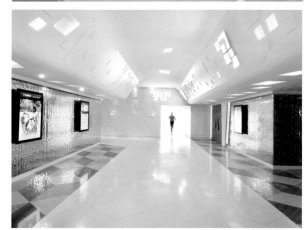

HOSPITALITY
餐饮休闲

Greyhound Café (ifc, HK)
Greyhound 餐厅（中国香港国际金融中心）

Greyhound Café (Harbour City, HK)
Greyhound 餐厅（中国香港尖沙咀海港城）

Crystal Jade La Mian Xiao Long Bao (ifc, HK)
翡翠拉面小笼包餐厅（中国香港国际金融中心）

Crystal Jade La Mian Xiao Long Bao (LAB, HK)
翡翠拉面小笼包餐厅（中国香港LAB）

Greyhound Café
Greyhound Café 餐厅

ifc, Hong Kong, China
中国香港国际金融中心

Area : 201.97m²
Completion Year : 2011
Design : Alex Choi, Ian Leung
Photographer : Bobby Wu

From Bangkok which the city has absorbed international elements, it is the first time for this famed fusion Thai restaurant to pick up its recipes and to spread their taste outside its origin. The first stop is Hong Kong. Greyhound is not only famous for the dishes, but also the fashion. The classic fusion style - Industrial chic - with a bit twist of signature Thai colour completed the outline of the café. The café is located in one of the classiest shopping mall and business centre in central where the beautiful sea view lies alongside. Is it a waste to trap all customers in a dark and moody room? The answer is definitely yes to the designer, therein lies the answer as to why the designer discarded bricks and walls and created an open seating area that blurring the boundaries between inside and outside. Together with the wall mural all over the space, continue the café's soul: be funky, be funny.

颇负盛名的灰狗泰国餐厅来自于国际化都市曼谷,这是它位于泰国之外的第一家门店,香港是它的第一站。灰狗不仅以其菜品闻名,它的时尚品味也是原因之一。经典的混搭风融合潮流的工业风,再加上泰式经典颜色创造了这个时尚的餐厅。该餐厅位于最大的购物商场和最大的商圈之中,坐拥无敌海景。将顾客禁锢在暗沉而忧郁的房间是不是一种浪费?对于设计师来说是的。所以他们摈弃了砖墙,代之以模糊室内外空间的开阔座位区。墙面的壁画生动地呈现咖啡店的灵魂 玩世不恭,趣味盎然!

2011 | Greyhound Café (ifc, HK)

| 2011 | Greyhound Café (ifc, HK) |

Greyhound Café

Greyhound Café 餐厅

Harbour City, Hong Kong, China
中国香港海港城

Area : 2,691m²
Completion Year : 2012
Design : Alex Choi, Ian Leung
Photographer : Bobby Wu

2012 | Greyhound Café (Harbour city, HK)

Located in the Harbour City where the mall was embraced by the well-noted Victoria Harbour, the café was mockingly sited in the inner area with no sea view. While the competitors have their own huge outdoor platform with superb view, "Why can't we make one by ourselves?" So, designer played with the space and perception and made a unique platform for the café. Then, they are even now! Continue the industrial chic style, the café was once again showing its charm of fusion and the attitude of funky and funny. In review of its first café in Hong Kong ifc, the wooden element was a new member to the interior. The intention was to bring harmony in the family-base shopping mall and its visitors. Celebrities were constantly found in the mall, and they also like visiting the café, so all paparazzi was invited by the mural painters and they were stuck at every window. This sense of humor will trigger the laughter but never annoyance. And the space with so many energetic elements that makes the interior so striking.

咖啡店所在的海港城被闻名的维多利亚港环抱，而从店内却看不见海景，这可能有一点讽刺，其他的竞争者坐拥大片室外露台，纵观无敌海景，"我们为什么不能自己制造出个海景呢？"。于是设计师为咖啡店量身定做了一个独特的观景台。这样就公平了！咖啡店延续了潮流工业风格，散发出一股混搭魅力，同时标榜出玩世不恭却趣味盎然的时尚态度。不同于第一家品牌店，这次店内设计引入了木材元素，这为家庭氛围浓重的购物商场增添了一抹和谐感。许多名人常来此地购物，他们也喜欢来这家咖啡店，所以壁画艺术家在墙面绘制了正在四处窥探的狗仔队。这恰如其分的幽默总会博人一笑却并不恼人。生气盎然的空间更是令人惊喜不断。

2012 | Greyhound Café (Harbour city, HK)

2012 | Greyhound Café (Harbour city, HK)

Crystal Jade La Mian Xiao Long Bao

翡翠拉面小笼包餐厅

ifc, Hong Kong, China
中国香港国际金融中心

Area : 3,372m²
Completion Year : 2011
Design : Alex Choi, Eric Young
Graphic Designer : Kit Siu
Photographer : Bobby Wu

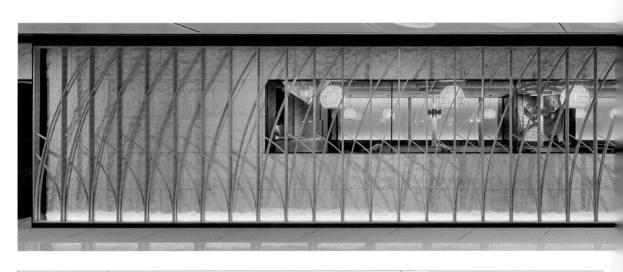

2011 | Crystal Jade La Mian Xiao Long Bao(ifc,HK)

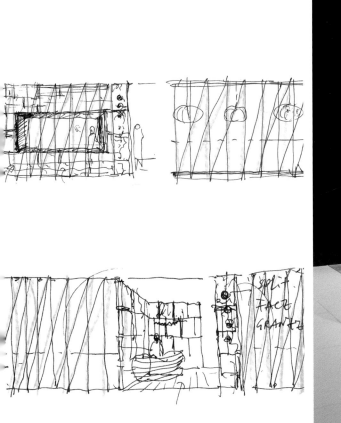

Extracting from the hero dish of the restaurant - Xiao Long Bao, bamboo baskets transformed into different forms in both abstract and physical way. Mix & Match with bamboo element, the facade was an art piece of traditional handcrafting. The woven bamboo trees spread their branches and crowns up to the top of the restaurant which re-create the scene of nature with a taste of art, and the use of earth tone increase the welcoming ambience through out the space. For another star dish in the restaurant is La Mian. The Japanese Artist, Studio Swanda Design, was invited to create art pieces named "Dancing Noodle". The "Dancing Noodle" was inspired by the making of La Mian. Alongside the dining area, a row of illuminated glass art-panel cannot be missed, the making of Xiao Long Bao also inspired designer in an innovated way. With reference to the circular cascade of ripples around the crown of Xiao Long Bao, the folding technique is imitated to the Wagami (Traditional Japanese paper) in the origination of the custom made wagami laminated glass panel. The "Xiao Long Bao" feature lighting gives emphasis to the pith of the restaurant. The well blend of materials usage creates layer by layer from in and out, the restaurant is full of texture with the diversity of material. More than simply a place for causal dinning, the restaurant is designed to evoke the art of traditional Chinese cuisine. Customers will more appreciate with the foods' origin and their story behind.

灵感来自于店里的招牌菜品——小笼包中，竹子被转化成了或实或虚的不同形态。店的外观是由竹元素混搭而成的传统手工制品。编织而成的竹树"枝桠"伸展覆盖了店面外观，颇具艺术感，另类地诠释了自然，大地色的使用增添了空间的友好氛围。日本艺术家Studio Swanda Design受邀创作了名为"舞动面条"的艺术品，创作灵感来自于店内的另一道名菜的制作过程：拉面。沿着就餐区布置的一排发亮的玻璃艺术面板不容错过。玻璃艺术面板也从小笼包上提取了灵感，日本传统纸张Wagami上折出小笼包的褶皱，最后变成这特别定制的艺术板。小笼包形状的灯盏再一次申明了餐厅的主题。从内到外不同材质的混合使用赋予餐厅多种不同质感。这不仅仅是一个就餐的地方，这是对中国传统美食文化的发扬。除了美食，顾客对于美食的历史渊源以及它们背后的故事也是饶有兴趣的。

Crystal Jade La Mian Xiao Long Bao
翡翠拉面小笼包餐厅

LAB, Hong Kong, China
中国香港 LAB

Area : 3,372m²
Completion Year : 2011
Design : Alex Choi, Eric Young
Graphic Designer : Kit Siu
Photographer : Bobby Wu

Located in the dense office worker habitat in Hong Kong, this Crystal Jade La Mian Xiao Long Bao has set an iconic 10m long semi-transparent magic mirror facade along the hustle and bustle pathway to draw some attention from hectic paces, whether customers and passersby can find their shadows through the obscure glass. At the end of the pathway next to the public entrance of the mall, there is a special display that showcasing the making-of of their signature cuisine dishes through the hazy glass — Xiao Long bao (Soup dumpling) and La Main (Noodle). In order to express the core concept of "handmade" in the interior design, the feature light which made of antique brass plate with hand-made nail mark was specially designed for the restaurant. While the shape of the restaurant is a fully enclosed rectangular box, designers has expanded the space through geometric and stereoscopic windows which was simply made by the mix and match of anodized stainless steel and mirror. To continue the art piece practice in the last concept restaurant, the cashier in this restaurant was touched up with the famous weaving method of bamboo baskets that always served with xiao Long bao. With all elements in this area, it has shown the oriental element integrated into the dining environment, people can get away from the hustle and bustle working life during their dining.

位于香港密集的上班族栖息地，这个翡翠拉面小笼包餐厅沿着熙熙攘攘的道路设置了一个标志性的10m长的半透明魔镜外立面，吸引匆匆而过的人群的注意力，不管是顾客还是路人都可以透过模糊的玻璃找到自己的影子。路的终点靠近购物中心公共入口，这儿有一个特别的展示，透过模糊的玻璃展现着他们的招牌食物的烹饪制作：小笼包（汤饺子）和拉面（面条）。为了在室内设计中表达核心理念"手工"，古黄铜板与手工钉标记制成的特色灯是专为餐厅而设计的。由于餐厅的形状是一个完全封闭的矩形框，设计师通过几何立体窗户（由阳极电镀不锈钢和镜面的混合与匹配简单制作而成）扩展了空间。为了在最后的概念餐厅延续艺术品实践，这个餐厅的收银台旁是用著名方法编织而成的竹篮——里面总是装着小笼包。这个区域的所有元素都显示了将东方元素融入到就餐环境，人们在用餐时可以远离喧嚣的工作生活。

RETAIL
零售

3 Mobile flagship store
3 移动电话旗舰店

3 Mobile chain store
3 移动电话连锁店

Hang Seng Bank retail branch revamp
恒生银行分银行形象改造

Garden of Eden
伊甸园

Leo Optic eyewear
护视光学眼镜店

TAG Heuer concept store
TAG Heuer 概念店

Moment fashion watch concept store
Moment 时尚手表概念店

ARTĒ Madrid concept store
艾尔蒂概念店

Watson's Wine Cellar chain store
屈臣氏酒窖连锁店

Fook Ming Tong Tea shop
福茗堂茶庄

Arome bakery room chain
东海堂连锁店

Dymocks book store
Dymocks 书店

3 Mobile flagship store

3 移动电话旗舰店

Hong Kong, China
中国香港

Area : 102.19m²
Completion Year : 2010
Design : Alex Choi, Yan Yik Lun
Photographer : Bobby Wu

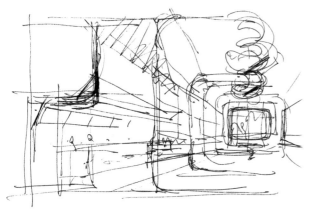

Fibre optics has been pushing to the front in the telecom industry this era, but fibre optics is always intangible. In order to modify this phenomenon and push it a next step forward, visualizing this technology became the main theme and mission of this 3 flagship store. The complex usage of white and grey enhances the depth of the space and creates an infinite perspective in a limited space. On the other hand the strips of mixing white and grey form a speed tunnel, customers immediately feel the sense of speed by setting their feet in the shop, the speed are also shown on the partition between the VIP sections. The "bushes" like display panel makes the theme idea extremely tangible to the whole design concept; A long row of "fibre optics" bushes leads the way through out the space, along with the hero products, customers can enjoy the adventure of exploring the latest digital products in the fibre bushes, and with the technology support, it has expanded a true interactive.

在这个时代,电讯业界一向对光纤推崇备至,但光纤往往都是难以捉摸的。为了把这个风行的技术推到更前线,光纤速度形象化就是这间3移动电话旗舰店的任务和主题。灰和白的运用提升了空间的深度,使整个空间无限地延伸。而且互相交接的灰白条子塑造了一条隧道,每当客人步入店内,便可以实时感受到光纤的速度,而VIP区的间板也体现出速度感。店的正中央竖立的"光纤丛林"展示柜,完全表达了整个设计理念;一排长长的光纤丛林加上主题产品贯穿了整个空间,顾客游走于丛林当中探索顶尖的电子产品,加上高科技的配合,达到真正的互动。

2010 | 3 Mobile flagship store

3 Mobile chain store
3 移动电话连锁店

Hong Kong, China
中国香港

Area : 102.19m²
Completion Year : 2010
Design : Alex Choi, Yan Yik Lun
Photographer : Bobby Wu

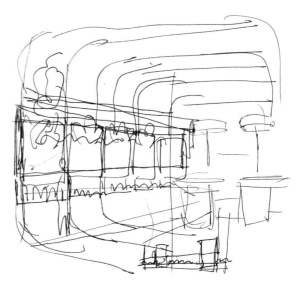

In this competitive industry, those competitors are only neck and neck in gaining the greatest market share. So, like all business men say: time is everything. While people are still exciting about the launch of the 3 flagship store in Central, the roll-out of the new image was undergoing at the same time. Although these shops are chain stores and under the same design concept, with the geographical varieties, the designers also tailor made for each shop which each of them has its own character. Getting rid of repeating the same design, the designers have given a different definition to chain store. So, like many of professional designers say: keeping the sight excitement is crucial.

这个行业竞争非常激烈，竞争对手并驾齐驱地抢夺着最大的市场份额。正如所有商人所强调的：时间就是一切。当人们还在为中心三家旗舰店的开张欢呼雀跃的时候，设计师已经在规划新的店铺形象。尽管是连锁店，有着相同的设计理念，但这些商店自身有着不同的地理特征，设计师便对每家商店进行量身设计，而不是都采用雷同的设计，从而给予了连锁店门不同的特性。正如许多专业设计师所说：保持视觉上的兴奋感至关重要。

| 2010 | 3 Mobile chain store |

Hang Seng Bank retail branch revamp
恒生银行分行形象改造

Hong Kong, China
中国香港

Completion Year : 2011 - 2013
Design : Alex Choi, Yan Yik Lun
Photographer : Bobby Wu, Apollo So

| 2011-2013 | Hang Seng Bank retail branch revamp |

In the year of 2012, the foremost branded local bank, Hang Seng Bank invited us to its revolutionary retail branch revamp. As a local designer, it is a great honor to take this challenge. This is a significant milestone for both ACD and the client. As the bank has a solid foundation of their steady and prudent financial services, designers must give mature consideration to all aspects of their social image. The key element of the brand was evolved such as 3D visualization of the Chinese ancient coin, focal usage of green colour and enlargement of pixel to enhance the bank's brand presence. Taking account to the operational needs, an airy, transparent and mobility design is the main theme of the mass banking area. The revolutionary design got rid of the traditional boxy design which was always adopted in bank design and with the touch of cosmetic treatment on the customer services cubicles, customers will feel more comfortable in the semi-open counter while privacy is retained. To emphasis the market leading role to the customers, the environment of prestige is tasteful and contemporary. Different layers of material such as tinted stainless steel, frosted tinted mirror, fabric texture wall paper delivered a semi glamorous style. The bold premium hotel environment of prestige reinforces the design in both function and presence and the enhancement can help to attract the affluent customers and also create a modern image of the bank. The refreshment of the bank brings fresh energy to both customers and staff. For examples, the "ribbon" design smoothes the flow of the area and the illuminated feature divider of Automatic Banking Center give a full sense of secure to customers etc. This fresh environment will encourage people to fast pace and enhance the efficiency of work in delightful atmosphere. Besides the street level branch, there are many convenience points in MTR stations such as automatic machines area and ATM which are in the high traffic location. In other words, they are the best advertisements in the public to promote the bank's new image. Get rid of the boxy design which is constrained by MTR's foot print and bring amazing effect with the new revolutionary design. Zigzag layout arrangement was applied in the rigid layout arrangement while stay respect to station condition but also pay attention to the linear flow of the traffic. The streamline canopy is inherited from the "ribbon" to soften the space. With the feature dividers, customers feel like under the umbrella of the bank.

2012年，当地最重要的银行品牌恒生银行邀请我们参与其革命性的零售支行改造。作为一个当地的设计师，他很荣幸能接受这个挑战。这对ACD和客户都是一个重要的里程碑。因此银行有一个坚实的基础，稳定而审慎的金融服务，设计师必须成熟地考虑到他们社会形象的所有方面。品牌的主要元素进化了，如中国古代钱币的3D可视化、绿色的集中使用和像素的增大，提高了银行的品牌形象。考虑到操作的需要，一种空灵、透明和机动性的设计成为大规模银行区的主要主题。革命性的设计摆脱了在银行设计中总是采用的传统的四四方方设计。对客户服务小隔间进行"整形治疗"，在顾客隐私被保留的同时，使顾客在半开的柜台感觉更舒服。为了向消费者突出市场的主导地位，气派的环境是当代雅致风格。不同层的材料如有色不锈钢、磨砂有色镜、织物纹理墙纸，呈现出半迷人的风格。气派而大胆的豪华酒店般环境加强了功能和形象设计，这种增强有助于吸引富有的客户，并建立现代银行形象。这个银行的革新给客户和员工带来新鲜的活力。例如，"丝带"设计减缓了区域的流动，自动银行中心带有照明功能的分频器给予客户充分意义上的安全等。在这个新的环境下，令人愉快的气氛会鼓励人们快节奏以及提高工作的效率。除了街道路口，在地铁站内有许多便利点，如在高流量位置的自动机器区和ATM。换句话说，它们是公共场合中促进银行新形象的最好广告。摆脱受到地铁脚印限制而形成的四四方方的设计，同这次革新一起带来惊人的影响。锯齿形布局代替僵直的布局安排，同时保持对站台条件的尊重，还要注重交通的线性流量。"丝带"取代了流线型的顶盖，增添了空间的柔软感。由于功能分隔器的存在，客户感觉像在银行的保护伞下。

| 2011-2013 | Hang Seng Bank retail branch revamp |

2011-2013 | Hang Seng Bank retail branch revamp

| 2011-2013 | Hang Seng Bank retail branch revamp |

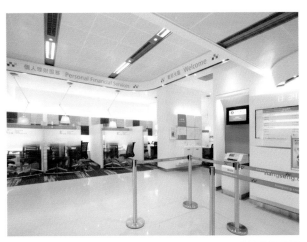

| 2011-2013 | Hang Seng Bank retail branch revamp |

| 2011-2013 | Hang Seng Bank retail branch revamp |

Garden of Eden
伊甸园

Hysan Place, Hong Kong, China
中国香港希慎广场

Area : 2,248.25m²
Completion Year : 2012
Design : Alex Choi, Ian Leung, Kim Ho
Graphic Designer : Kit Siu
Photographer : Bobby Wu

To go with the theme of intimacy & lingerie concept floor in Hysan Place, the very first thing came to the designer's mind was leaves – the first underthings on earth that wore by Eve. And when the idea came across, the scenario of Garden of Eden was set in the right place. This concept floor is of a diverse nature from a theme park design, design elements are ingeniously sprinkled through out the whole space: leave feature ceiling, leave pattern floor and feature pendent light which is the symbol of apple – fruit from the tree of knowledge. Stepping in the restroom, the snake skin wall and patch work floor will lead the way to the basin, which is shaped into a sliced apple. All components are transformed and presented in an abstract way that makes this floor sophisticated and special, and variousness of shops will be the best accessories of the whole floor.

为了与希慎广场的概念楼层"亲密内衣"主题相符，设计师想到的第一个事物是叶子——地球上第一个女子夏娃穿着的内衣裤。当这个想法出现时，伊甸园的场景以正确的方式进行了布置。这个概念楼与主题公园的设计性质不同。设计元素巧妙地布置在整个空间：树叶形状的天花板、树叶图案的地板以及悬垂的苹果形状的灯——来自智慧树上的水果。步入洗手间，蛇皮纹的墙壁和斑点状工作地板通向呈苹果片状的洗手盆。所有的组件都以一种抽象的方式进行转变和呈现，使得这个楼层精致而特别，而多样化的商店将是整个楼层最好的配件。

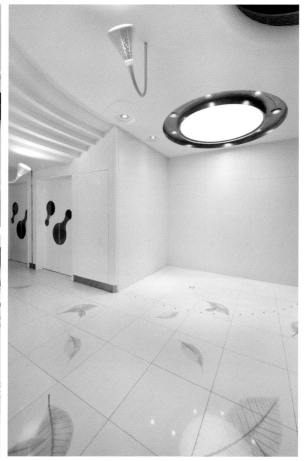

2012 | Garden of Eden

Leo Optic eyewear
护视光学眼镜店

ifc, Hong Kong, China
中国香港国际金融中心商场

Area : 38.74m²
Completion Year : 2011
Design : Alex Choi, Ian Leung
Photographer : Bobby Wu

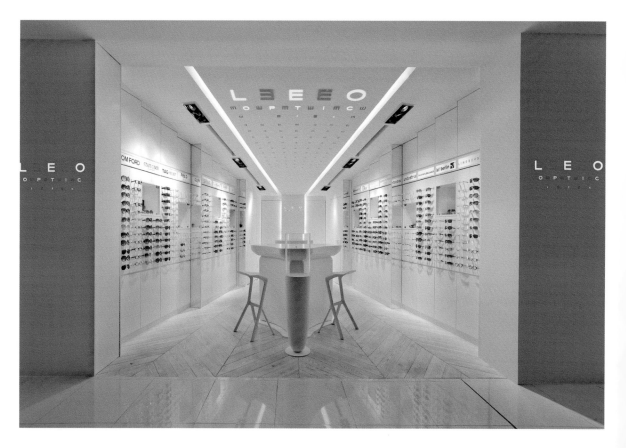

2011 | Leo Optic eyewear

To enhance the professional status of eyewear shop in one of the most high class shopping malls in Hong Kong, the designer is designated to give a totally new renovation of the shop - in and out. Extricate itself from a conventional, ordinary and unstriking shop to a professional and high end eyewear shop. Owing to its size limitations, simplicity form set the main tone of the theme. Generally apply white on the shop to ferment the clinical atmosphere and also to lay emphasis on its professional status. The accent is also shown on the ceiling which slopes to the focal point of the shop, the eye chart is filled above. Exaggerating the one point perspective in the long space and elongated the shop with the symmetrical display units and inclination of the setting out. The directional forms of the refined strip oak wood flooring stretch to the end of thevanishing point. Besides putting the logo at the middle top of the portal, designer discards the traditional way and incorporates the logo onto the ceiling from the shop front to the focal point. The wall-hanging design leaves sufficient space for the walkway and with the open setting of the display unit, the multitude of glasses decorated the shop and outstands themselves from the light trough highlight. While everything in its right place, onlookers will be all fooled by the illustration of the depth.

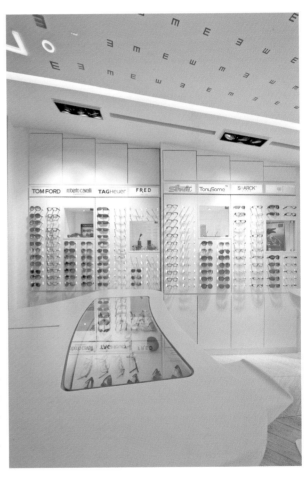

为了提高位于香港最高档购物中心的眼镜店的专业形象，设计师被委任给予商店内外一个全新的改造。将原本传统、普通、没有吸引力的店转变成专业、高端的眼镜店。由于它的面积限制，将简单的形式定为空间主调。通常，应用白色发酵冷静的气氛，以及强调其专业形象。天花板上的重心倾斜到商店的中心，视力检查图填充在上面。放大长形空间里的一个视角点，用对称的陈设单元和倾斜的排列从视觉上延伸商店面积。定向形式的精制带状橡木地板延伸到最后的消失点。除了把商店标识放在门面的中间顶部，设计师抛弃了传统的方式，从店面到中心的天花板上印着商店标识。壁挂设计为通道留下了充分的空间，打开陈设单元，众多的眼镜装饰着商店，在光槽的强光中突显出来。同时，一切都在合适的位置，由于店内的布局，空间在视觉上会形成延伸。

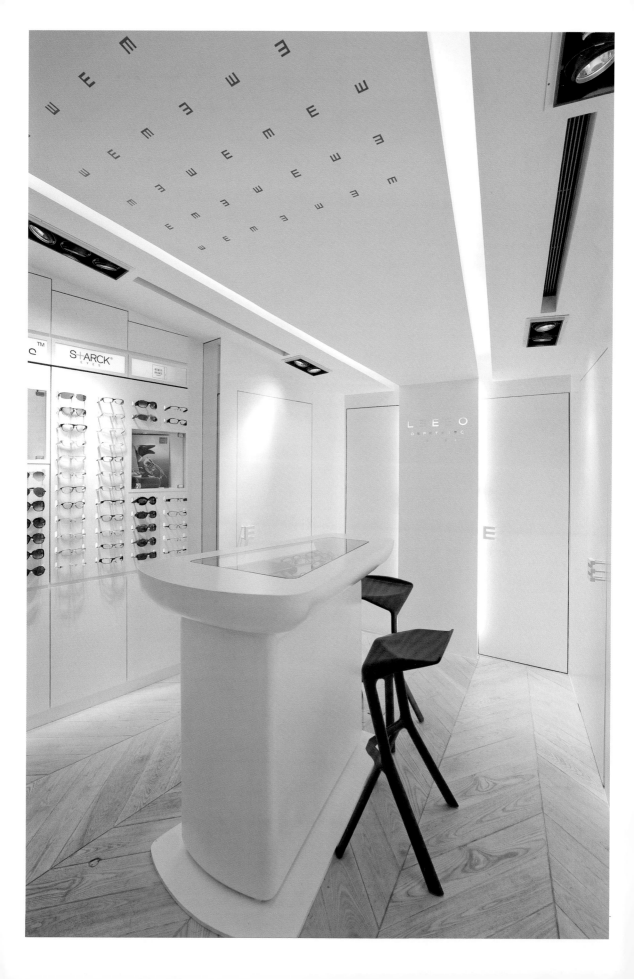

TAG Heuer concept store

TAG Heuer 概念店

ifc, Hong Kong, China
中国香港国际金融中心商场

Area : 37.16m²
Completion Year : 2012
Design : Alex Choi, Kim Ho
Photographer : Bobby Wu

Time is interesting in the universe. It is a concept, a rule, a unit, etc. Time can be anything. TAG Heuer has a global recognition of its delicate and luxury watches. This 37.16m² space articulates the identity of the brand with organic curves, which the grooves can be the best modifier to embody the form of time. The continuing window display connects the shop from the inside to the outside, and the end point of the display units are vanished by design. This is the idea to prolong the rhythm through out the shop and it also matches with the concept "time never stops". The LED spot lights along the grooves cast light on the displays and fill the shops with multiple layers. The anodized aluminum cladding wall panel with indistinct seashell patterns and the patterns appear arbitrarily under the pastel shades of lighting effect. It is not bold but the impression is stunning when the pattern shows up in various angles. This unique finish is a perfect match of the brand's subtlety.

时间在这个宇宙里是多么有趣，它是一个概念、一条规则、一个单元，时间可以是任何东西。TAG Heuer 精致奢华的手表是世界认同的。在这个 37.16m² 的空间里，有机曲线强调了该品牌的特性。曲线最能诠释时间的形态。贯穿店内的玻璃展柜将室内外空间联为一体，巧妙的设计制造出一种空间的无限感。这延伸了店内的空间，同时与"时间永不停歇"不谋而合。

LED 灯沿着曲线照射着展柜，并营造出多层次的空间。氧化处理的铝质墙板融入了模糊的贝壳图案，图案在水彩色的光影之中若隐若现，不同角度的光照射到它时产生的效果令人惊叹，虽然初看并不显眼。这独特的细节处理与品牌的精致品质可以说是相辅相成。

Moments fashion watch concept store

Moments 时尚手表概念店

ifc, Hong Kong, China
中国香港国际金融中心商场

Area : 61.3m²
Completion Year : 2011
Design : Alex Choi, Yan Yik Lun
Photographer : Bobby Wu

2011 | Moments fashion watch concept store

Moments is a sub-brand that derived from a renowned local fashion watch shop. Different from its mother brand, Moments' customer profile is young professionals, hence semi-luxury fashion brands timepieces can be found in the shop. Keeping the concept of "time is infinite", designers carved out a new spatial identity without losing the soul of the classy. The highlight of orange accent at the feature ceiling enhanced its vibrant and avant-grade ambience.

This shop is located in one of the hippest shopping mall in Asia, new materials were selectively applied in and out, such as the lighten luminous counter top which is made by concrete blocks with embedded webbed fiber optic cables, and the sculptured aluminum panel at the shop front. To create a new kind experience of shopping, watches are set out on islands which were built into the interior false pillars. As the watches are displayed at our eye level, customers no longer need to make bow while browsing around.

Moments 是一个本地著名表店的附属品牌。不同于主品牌的客户定位，Moments 瞄准了年轻白领，采取了中高端定位。设计师探索了一种全新的空间设计手法，却没有丧失经典的内涵，保留了"时间无限"的品牌概念。天花板的橘色十分吸引眼球，提升了空间的灵动前卫感。

这家店位于亚洲最热闹的几条购物街之一，内外全部采用了最新材料，像是使用混泥土块嵌入网状光纤电缆制作的发光柜台，店前的雕刻铝版。为了制造购物新体验，嵌入假式台柱的展台取代了传统展柜让顾客不用弯腰也能水平直视产品。

Moments fashion watch concept store

ARTĒ Madrid concept shop

艾尔蒂概念店

ifc, Hong Kong, China
中国香港国际金融中心商场

Area : 28.7m²
Completion Year : 2012
Design : Alex Choi, Ian Leung
Photographer : Bobby Wu

2011 | ARTĒ Madrid concept store

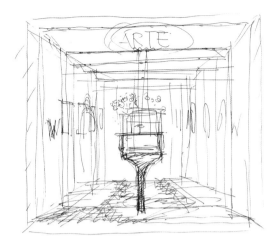

This fashion jewellery brand was brought from Madrid, Spain, a city that full of exotic allure. Elegant and feminine were always used to describe the brand but it not the whole story. Finally, there is a chance for the brand to speak for itself. A be spoken design is always emphasized in the designers philosophy. This time, besides the general description of its elegant, the designers demonstrated the other side behind luxury by the element of chic, unisex, energetic and open. Differ from a rather concealed and mysterious ambience that most of the jewellery shops may convey, an open shop front had seem more likely to show the brand's vibrant and glamorous character. The dancing display niches on the wall and imitation of geometric shape of glamorous gemstones have broken the boxy image of all display unites and bring in the vividness to the shop. Extract from the craftsmanship of sculpting and finishing skills, the ingenious geometric lines was carved from the wall to the ceiling. With the brightening tone of the shop, there was an air of excitement at the shopping time.

该时尚珠宝店来自西班牙马德里，一个充满异国情调魅力的城市。优雅的、女性气质的常常被用来形容这个品牌，但这并不是它的全部。最后，这是个机会，品牌为自己代言。可以用言语描述的设计总是突出在设计师的设计哲学里。这次，除了对它的优雅作大体的呈现，设计师以其别致、不分男女、充满活力和开放的元素，展示它奢华背后的另一面。不像大多数珠宝店传递的相当隐匿而神秘的气氛，这个开放的店面更多地展示了这个品牌充满活力、富有魅力的特征。墙上的舞蹈展示壁龛对几何形状的迷人宝石的模仿，打破了陈设单元四四方方的形象，给商店带来了活力。吸取工艺雕刻和装饰技巧的精华，从墙上到天花板雕刻巧妙的几何线条。由于商店明亮的色调，顾客在购物的期间气氛显得有些兴奋。

Watson's Wine Cellar chain store
屈臣氏酒窖连锁店

Hong Kong, China
中国香港

Completion Year : 2012
Design : Alex Choi
Photographer : Bobby Wu

Years before, wine tasting is an extravagant habit that favoured fat cats. As the wine tasting culture become more and more popular, besides spending billions of dollars on flying all the way across the globe to visit those wine cellars, now people have another choice. Without a doubt that traditional wine cellar has its allure that can hardly be replaced, urban wine cellars had given people flexibility to have high quality wine in spitting distance. Although Watson's Wine Cellar is a chain store, the designer wanted to emphasize the expertise. Extract from signature mark of wine cellar, arc ceiling was applied to every shop; also the shop front portal was created by a bunch of wine bottles which gives a strong signature to this "wine cellar" like store. The well-mixed of steel and brass multiplied the effect.

许多年前，品酒是一个奢侈的习惯，是有钱有势人的专利。随着品酒文化越来越流行，除了花费数十亿美元用于飞行世界各地访问那些酒窖，现在，人们有了另一个选择。毫无疑问，传统酒窖有它难以取代的魅力，而城市酒窖让人们在近距离灵活地品尝高质量酒。虽然屈臣氏的酒窖是一个连锁店，但设计师仍想强调其专业技术。弧形从酒窖的标志中提取而来，弧形天花板应用到每一家店；店面大门是由一堆酒瓶塑造而成，给予这个"酒窖"强有力的签名。钢和铜的混合增强视觉效果。

Fook Ming Tong Tea shop
福茗堂茶庄

ifc, Hong Kong, China
中国香港国际金融中心商场

Area : 58.90m²
Completion Year : 2011
Design : Alex Choi, Stephanie Leung
Graphic Designer : Kit Siu
Photographer : Bobby Wu

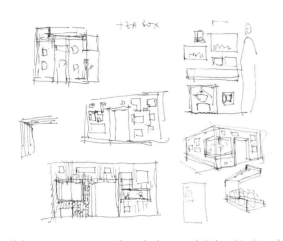

Chinese tea, more than being a drink, it is also recognized as a type of Chinese medicine for health and beauty and wisdom of ancient sages. To continue the tea culture and carry on to the next generation, who lives in "la dolce vita", a well-blend with modern and traditional tea house is brought to the table. Located in the most luxury shopping mall in Hong Kong, the brand new look of Fook Ming Tong seeps out the real freshness. Extracting the concept from traditional Chinese hand crafted art, the shop is designed as a delicate ancient jewellery case which it symbolizes the diversity. In this case, only valuable tea leaves is collected. On the tea sample feature wall, customers can explore more about the tea while steeping themselves in the great scent of tea. Instead of taste, the spirit of tea tasting is about the enjoyment and artistic conception. This spiritual concept is also shared with Chinese water color painting. Just take a distant view on the facade, a stunning laser engraved Chinese water color brush painting interprets the abstract concept in the way you can taste.

中国茶叶，不仅仅是一种饮品，它也被认为是一种对健康与美丽有益及增长古代圣贤智慧的中药。为了延续茶文化并传承给过着"甜蜜生活"的下一代，一个现代和传统完美结合的茶室呈现出来。位于香港最豪华的购物中心，福茗堂的品牌新形象渗透出真正的新鲜感。从中国传统的手工艺术中提取概念，商店被设计成一个精致的古代首饰盒，象征着多样性。在这种情况下，只有有价值的茶叶才被收集。在茶样特色墙上，客户可以探索更多关于茶的内容，同时把自己浸泡在极好的茶香里。品茶的精髓不是味道，而是口感享受和意境。这种精髓概念也用于中国水彩画。远观外立面，一个惊人的激光雕刻的中国水彩毛笔画诠释着茶道的抽象概念。

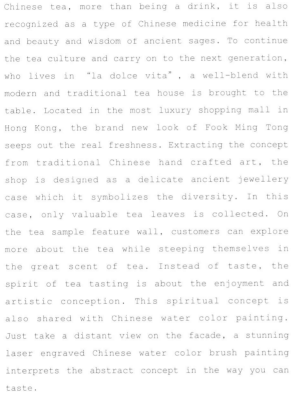

Arome bakery room chain
东海堂连锁饼店

Hong Kong, China
中国香港

Completion Year : 2011 - 2013
Design : Alex Choi, Clarissa Ng, Ian Leung
Photographer : Bobby Wu

2011-2013 | Arome bakery room chain

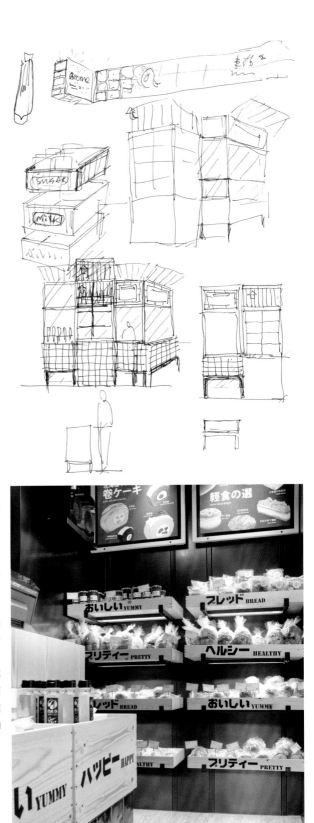

Everything started from the orientation of the Arome Bakery room, Japan, where the dainty pastry is famous in worldwide. With the well mix of western and Japanese ingredients and technique, every pastry is the master piece of the pastry maker. To catch up with the fame of the dainty pastry, the interiors combine delicate elements with western influences and make it simply western orient. Extract from the Japanese traditional handcrafting and elegance of western style, the elements are with high design efficiency for the future roll-outs and they are selectively melted in the space. Serving as a focal point, the western second hand antique lights featuring the ash wood hand craft menu board and the abandoned plywood box completes a scene devoted to the senses. The home made bakery counter boasts with England antique jasper handmade glaze title enhance the elegance of the bakery room. For head to toe, the blends of 2 cultures are just systematically situated in the space and cooked up for pastry lovers to taste.

一切源于东海堂饼店的发源地——日本；当地闻名的精美巧手糕点，是糕点师把日式及西式的糕点精髓融合，以巧手制成的杰作。为了把这个特征呈现出来，饼店的新形象揉合了西方的色彩，加上精致的装饰元素，令东海堂由内到外展现出东洋魅力。利用西方的贵气和日本传统手工艺的特色融入整个空间，亮点是以英国墨绿色手工磁砖包裹着的自家制面包房，加上西式的古董灯饰衬托着手制原木餐牌，使整个东洋观感更高贵完整。从上而下，层次分明的东西文化风格，正好给顾客品尝其中的东洋之味。

2011-2013 | Arome bakery room chain

2011-2013 | Arome bakery room chain

Dymocks book store
Dymocks 书店

ifc, Hong Kong, China
中国香港国际金融中心商场

Area : 455.22m²
Completion Year : 2011
Design : Alex Choi
Photographer : Bobby Wu

Dymocks is a leading Australia-based bookseller in the Asia Pacific region, which has 70 stores in Australia and Hong Kong,China. As a time-honored brand, a global design guideline has been implemented usually. When Dymocks comes to Hong Kong, they have taken a big step to accept the distinctive interior design. The whole picture of the interior design creates a variety of places; places to ponder, to linger, or to sit and simply immerse oneself in a story. The endless bookcases along the facade and throughout the ceiling and a reading circuit seem to be trying to match the concept of endlessness of the universe, you can feel as free as if you were wandering in a forest until stumble upon an interesting book by accident. Book is a kind of medium to enlighten people's mind, so the designer uses different kind of feature lamps surrounding the centre of the book store to represent plenty of ideas. For book lovers, it is an ideal location where they can conjure up the boundless possibilities.

Dymocks 是一个以澳大利亚为基地的亚太地区顶级书商,在澳大利亚和中国香港拥有 70 家店。作为一个历史悠久的品牌,通常实施全球设计指导准则。而当 Dymocks 来到香港,他们迈出了一大步去接受独特的室内设计。整个室内设计图有着各种各样的地方,这些地方用于思考、逗留,或仅仅坐着沉浸在自己的故事里。沿着外立面和整个天花板的无限书架以及一个读书路线,似乎试图与宇宙的无边无际的概念相一致。在这,你会感到无拘无束,仿佛你是在一个森林里游荡,直到偶然发现一本有趣的书。书是启迪人们心灵的一种介质,所以设计师使用不同类型的特色灯围绕在书店的中心,以代表各种思想。对于图书爱好者,这是一个理想的地方,在这里他们可以变幻出无限的可能。

2011 | Dymocks book store

OFFICE
办公室

Royal Spirit office
Royal Spirit 办公室

The ROOM studio
The ROOM 工作室

Royal Spirit office
Royal Spirit 办公室

Hong Kong, China
中国香港

Area : 4,004.12m²
Completion Year : 2011
Design : Alex Choi
Photographer : Bobby Wu

| 2011 | Royal Spirit office |

The industrial re-revolution is happened in textile office located at a 70's factory Building which carries a long history. Back to the 70's, toys, textiles, watches and jewelleries, are the 4 king industries that had brought Hong Kong an economic summit. But time flies, the heroes have been forgotten like they have never existed. Born in 70's, designer's mission is to bring the classic back and raise the industrial re-revolution. This textile company needed a repackage service to hark back people's memory and meet the trend. Like a quantum leap effect: 2 extreme elements meet and cause a big bang. Primary material is constantly used in the space, like iron, steel, glass, brick etc. These materials maybe common to the world nowadays, and people also take them for granted. But the fact is, these materials are used to be great inventionsthat had stunned the world with their utility, and they are always playing their part by improving peoples living. These harmonious materials create a minimal ambience by only using black, white and grey as the theme colour of the space,keeping the area true and humble with no extra accessories when the natural materials are the best trimmings, and the use of dramatic lighting shines out the true beauty of the materials.

二次产业革命是发生在 70 年代一个有着悠久历史的厂房的纺织办公室里。追溯到 70 年代，玩具、纺织品、手表和珠宝，4 个首屈一指的产业带来了香港经济发展的巅峰。但是，时间飞逝，英雄已经被人遗忘，好像他们从来不曾存在过。出生在 70 年代，设计师的使命是重塑经典、进一步促进二次产业革命。这个纺织公司需要一个重新包装服务，以唤起人们的记忆和满足潮流发展的需要。像一个量子飞跃般效应：2 个极端元素相结合，产生巨大的反响。主要材料被重复运于空间，如铁、钢、玻璃、砖……这些材料也许在当今世界是常见的，人们也把它们视为理所当然。但事实是，这些材料都曾经是伟大的发明，它们的作用震惊了世界，它们总是在改善人们生活中起着重要作用。只使用黑色、白色和灰色为空间的主题颜色，这些和谐的材料创建了一个最小的气氛；自然材料就是最好配饰，不使用额外的配饰保持了该地区真实和谦卑感；并使用引人注目的灯光效果照射出材料真正的美感。

2011 | Royal Spirit office

2011 | Royal Spirit office

2011 | Royal Spirit office

The ROOM studio

The ROOM 工作室

Hong Kong, China
中国香港

Area : 410.5m²
Completion Year : 2011
Design : Alex Choi
Photographer : Kit Siu, Tat Wong

2011 | The ROOM studio

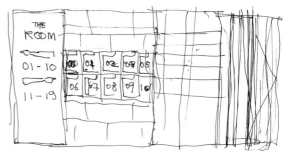

Each moment in history is a fleeting time, precious and unique. Back to 70's Hong Kong, the development of industries are prosperous, it brought Hong Kong an economic summit. Although time flies, the unique philosophy and its embodiment as studio for young generation open their eyes to true value of manufacturing and design industry in today community through a space of raw, simple and practical. Primary material is constantly used, like raw concrete wall in lobby, wooden mail box and pine wood toilet door and gypsum plasterboard as partition for every room. These back-to-basics material create a minimal atmosphere and keep the area as simplistic as 70s. Young designers, manufacturers and artists could utilize this practical space to continue their own manufacture, at the same time to create another boom in The ROOM.

历史上的每一刻都是瞬间，弥足珍贵。追溯到 70 年代的香港，工业发展繁荣，带来香港经济发展的巅峰。虽然时间飞逝，工作室的独特哲学和具体体现为年轻一代打开了眼界，通过原始、简单、实用的空间，展示它在当今社会制造和设计行业的真实价值。主要材料被不断使用，像在走廊的原始混凝土墙、木质邮箱与松木卫生间门、作为每个房间隔墙的石膏板。这些返朴归真的材料创建了一个最小的氛围，保持该区域 70 年代的简化风格。年轻的设计师、制造商和艺术家可以利用这个实用空间继续他们自己的生产，同时在 The Room 工作室里创造另一个繁荣。

| 2011 | The ROOM studio |

The ROOM studio | NO.118 - 119

| 2011 | The ROOM studio |

Acknowledgements

We would like to thank all the designers and companies who made significant contributions to the compilation of this book. Without them, this project would not have been possible. We would also like to thank many others whose names did not appear on the credits, but made specific input and support for the project from beginning to end.

Future Editions

If you would like to contribute to the next edition of Artpower, please email us your details to: artpower@artpower.com.cn